WITHDRAWN

271
Topics in Current Chemistry

Editorial Board:
V. Balzani · A. de Meijere · K. N. Houk · H. Kessler · J.-M. Lehn
S. V. Ley · S. L. Schreiber · J. Thiem · B. M. Trost · F. Vögtle
H. Yamamoto

Topics in Current Chemistry
Recently Published and Forthcoming Volumes

Bioactive Confirmation I
Volume Editor: Peters, T.
Vol. 272, 2007

Biomineralization II
Mineralization Using Synthetic Polymers and Templates
Volume Editor: Naka, K.
Vol. 271, 2007

Biomineralization I
Crystallization and Self-Organization Process
Volume Editor: Naka, K.
Vol. 270, 2007

Novel Optical Resolution Technologies
Volume Editors:
Sakai, K., Hirayama, N., Tamura, R.
Vol. 269, 2007

Atomistic Approaches in Modern Biology
From Quantum Chemistry
to Molecular Simulations
Volume Editor: Reiher, M.
Vol. 268, 2006

Glycopeptides and Glycoproteins
Synthesis, Structure, and Application
Volume Editor: Wittmann, V.
Vol. 267, 2006

Microwave Methods in Organic Synthesis
Volume Editors: Larhed, M., Olofsson, K.
Vol. 266, 2006

Supramolecular Chirality
Volume Editors: Crego-Calama, M., Reinhoudt, D. N.
Vol. 265, 2006

Radicals in Synthesis II
Complex Molecules
Volume Editor: Gansäuer, A.
Vol. 264, 2006

Radicals in Synthesis I
Methods and Mechanisms
Volume Editor: Gansäuer, A.
Vol. 263, 2006

Molecular Machines
Volume Editor: Kelly, T. R.
Vol. 262, 2006

Immobilisation of DNA on Chips II
Volume Editor: Wittmann, C.
Vol. 261, 2005

Immobilisation of DNA on Chips I
Volume Editor: Wittmann, C.
Vol. 260, 2005

Prebiotic Chemistry
From Simple Amphiphiles to Protocell Models
Volume Editor: Walde, P.
Vol. 259, 2005

Supramolecular Dye Chemistry
Volume Editor: Würthner, F.
Vol. 258, 2005

Molecular Wires
From Design to Properties
Volume Editor: De Cola, L.
Vol. 257, 2005

Low Molecular Mass Gelators
Design, Self-Assembly, Function
Volume Editor: Fages, F.
Vol. 256, 2005

Anion Sensing
Volume Editor: Stibor, I.
Vol. 255, 2005

Organic Solid State Reactions
Volume Editor: Toda, F.
Vol. 254, 2005

Biomineralization II

Mineralization Using Synthetic Polymers and Templates

Volume Editor: Kensuke Naka

With contributions by
H. Cölfen · K. Naka · T. Okamura · A. Onoda
K. Takahashi · N. Ueyama · H. Yamamoto · S.-H. Yu

 Springer

The series *Topics in Current Chemistry* presents critical reviews of the present and future trends in modern chemical research. The scope of coverage includes all areas of chemical science including the interfaces with related disciplines such as biology, medicine and materials science. The goal of each thematic volume is to give the nonspecialist reader, whether at the university or in industry, a comprehensive overview of an area where new insights are emerging that are of interest to a larger scientific audience.

As a rule, contributions are specially commissioned. The editors and publishers will, however, always be pleased to receive suggestions and supplementary information. Papers are accepted for *Topics in Current Chemistry* in English.

In references *Topics in Current Chemistry* is abbreviated Top Curr Chem and is cited as a journal.

Visit the TCC content at springerlink.com

ISSN 0340-1022
ISBN-10 3-540-46376-3 Springer Berlin Heidelberg New York
ISBN-13 978-3-540-46376-4 Springer Berlin Heidelberg New York
DOI 10.1007/978-3-540-46378-8

This work is subject to copyright. All rights are reserved, whether the whole or part of the material is concerned, specifically the rights of translation, reprinting, reuse of illustrations, recitation, broadcasting, reproduction on microfilm or in any other way, and storage in data banks. Duplication of this publication or parts thereof is permitted only under the provisions of the German Copyright Law of September 9, 1965, in its current version, and permission for use must always be obtained from Springer. Violations are liable for prosecution under the German Copyright Law.

Springer is a part of Springer Science+Business Media

springer.com

© Springer-Verlag Berlin Heidelberg 2007

The use of registered names, trademarks, etc. in this publication does not imply, even in the absence of a specific statement, that such names are exempt from the relevant protective laws and regulations and therefore free for general use.

Cover design: WMXDesign GmbH, Heidelberg
Typesetting and Production: LE-TEX Jelonek, Schmidt & Vöckler GbR, Leipzig

Printed on acid-free paper   02/3100 YL – 5 4 3 2 1 0

Volume Editor

Dr. Kensuke Naka

Kyoto University
Graduate School of Engineering
Department of Polymer Chemistry
Katsra, 615-8501 Kyoto, Japan
ken@chujo.synchem.kyoto-u.ac.jp

Editorial Board

Prof. Vincenzo Balzani

Dipartimento di Chimica „G. Ciamician"
University of Bologna
via Selmi 2
40126 Bologna, Italy
vincenzo.balzani@unibo.it

Prof. Dr. Armin de Meijere

Institut für Organische Chemie
der Georg-August-Universität
Tammanstr. 2
37077 Göttingen, Germany
ameijer1@uni-goettingen.de

Prof. Dr. Kendall N. Houk

University of California
Department of Chemistry and
Biochemistry
405 Hilgard Avenue
Los Angeles, CA 90024-1589
USA
houk@chem.ucla.edu

Prof. Dr. Horst Kessler

Institut für Organische Chemie
TU München
Lichtenbergstraße 4
86747 Garching, Germany
kessler@ch.tum.de

Prof. Jean-Marie Lehn

ISIS
8, allée Gaspard Monge
BP 70028
67083 Strasbourg Cedex, France
lehn@isis.u-strasbg.fr

Prof. Steven V. Ley

University Chemical Laboratory
Lensfield Road
Cambridge CB2 1EW
Great Britain
Svl1000@cus.cam.ac.uk

Prof. Stuart L. Schreiber

Chemical Laboratories
Harvard University
12 Oxford Street
Cambridge, MA 02138-2902
USA
sls@slsiris.harvard.edu

Prof. Dr. Joachim Thiem

Institut für Organische Chemie
Universität Hamburg
Martin-Luther-King-Platz 6
20146 Hamburg, Germany
thiem@chemie.uni-hamburg.de

Prof. Barry M. Trost

Department of Chemistry
Stanford University
Stanford, CA 94305-5080
USA
bmtrost@leland.stanford.edu

Prof. Dr. F. Vögtle

Kekulé-Institut für Organische Chemie
und Biochemie
der Universität Bonn
Gerhard-Domagk-Str. 1
53121 Bonn, Germany
voegtle@uni-bonn.de

Prof. Dr. Hisashi Yamamoto

Department of Chemistry
The University of Chicago
5735 South Ellis Avenue
Chicago, IL 60637
USA
yamamoto@uchicago.edu

Topics in Current Chemistry
Also Available Electronically

For all customers who have a standing order to Topics in Current Chemistry, we offer the electronic version via SpringerLink free of charge. Please contact your librarian who can receive a password or free access to the full articles by registering at:

springerlink.com

If you do not have a subscription, you can still view the tables of contents of the volumes and the abstract of each article by going to the SpringerLink Homepage, clicking on "Browse by Online Libraries", then "Chemical Sciences", and finally choose Topics in Current Chemistry.

You will find information about the

- Editorial Board
- Aims and Scope
- Instructions for Authors
- Sample Contribution

at springer.com using the search function.

Preface

In nature, biological organisms produce mineralized tissues such as bone, teeth, diatoms, and shells. Biomineralization is the sophisticated process of production of these inorganic minerals by living organisms. Construction of organic–inorganic hybrid materials with controlled mineralization analogous to those produced by nature has recently received much attention because it can aid in understanding the mechanisms of the biomineralization process and development of biomimetic materials processing. The biomineralization processes use aqueous solutions at temperatures below 100 °C and no toxic intermediates are produced in these systems. From a serious global environmental problem point of view, the development of processes inspired by biomineralization would offer valuable insights into material science and engineering to reduce energy consumption and environmental impact. One of the most challenging scientific problems is to gain greater insight into the molecular interactions occurring at the interface between the inorganic mineral and the macromolecular organic matrix. Model systems are often regarded as a straight-forward experimental approach toward biomimetic crystallization. Hierarchical architectures consisting of small building blocks of inorganic crystals are often found in biominerals. Studies of nanocrystal self-organization in solution systems would also be helpful for understanding biomineralization.

In these volumes, we focus on construction of organic–inorganic hybrid materials with controlled mineralization inspired by natural biomineralization. In the first volume, the reader will find contributions providing a basic scope of the mineralization process in aqueous solution. The first chapter by Marc Fricke and Dirk Volkmer introduce Mollusk shell formation, crystallization of $CaCO_3$ underneath monolayer via epitaxial and non-epitaxial growth. Hiroaki Imai describes hierarchically structured crystals formed through self-organization in solution systems. A wide variety of complex morphologies including fractals, dendrites, and self-similar structures are reviewed. Fluoratite-gelatine nanocomposites, reviewed by Rüdiger Kniep and Paul Simon, are suited to obtaining deeper insight into processes of self-organization, and can help to learn about essentials in the formation of inorganic–organic nanocomposites of biological relevance. There is considerable academic and commercial interest in the development of hydroxyapatite (HAp) bioceramics and HAp-loaded polymer for bone replacement. Kimiyasu Sato summarized inorganic–organic interfacial interactions in hydroxyapatite mineralization process. Biominerals are most often considered in either their more traditional roles as critical structural components of organisms. The biominerals are also important ion reservoirs for cellular function. It is becoming apparent that biominerals act as critical detoxification sinks within certain organisms. In the final chapter of the first volume, David Wright and his co-workers explain that the detoxification process make noble metal nanoparticles.

In the mineralized tissues, crystal morphology, size, and orientation are determined by local conditions and, in particular, the presence of "matrix" proteins or other macromolecules. Because the proteins that have been found to be associated with biominerals are usually highly acidic macromolecules, several water-soluble polyelectrolytes have been examined for the model of biomineralization in aqueous solution. The second volume will focus on controlled mineralization by synthetic templates. We start this volume with the latest advances in hydrophilic polymer controlled morphosynthesis and bio-inspired mineralization of crystals reviewed by Helmut Cölfen. He also summarizes classical crystallization pathways as well as non-classical nanoparticle mediated crystallization routes. These basic overviews will be very useful for the reader to understand this field. Shu-Hong Yu reviews controlled mineralization by synthetic additives and templates, not only with simple water-soluble polymers, but also artificial interfaces and matrixes including monolayers and synthetic polymer matrix. In the next chapter, I will introduce delayed action of polymeric additives as a new method for controlled mineralization by synthetic polymers. Finally, Norikazu Ueyama and his co-workers describe the detailed chemistry of the interface between the inorganic mineral and the macromolecular organic matrix by their designed polymer ligands. Coordination information on synthetic Ca carboxylate complexes is important for the elucidation of the Ca–O bond on the surface of Ca-based biominerals.

It was my great pleasure to invite leading international scientists to contribute to this issue and write excellent and detailed reviews of recent developments in the field of biomineralization. Although several books about biomineralization have been published, those books focused on the progress of biomineralization in biology and molecular biology. No book about biomineralization has focused on the viewpoint of constructing the bio-inspired organic–inorganic hybrid materials. The continuous cooperation of organic and polymer chemists with inorganic and biochemists for the field of biomineralization is desirable for discovering new concepts and method for producing composite materials and crystalline forms analogous to those produced by nature.

Kyoto, September 2006   Kensuke Naka

Contents

Bio-inspired Mineralization Using Hydrophilic Polymers
H. Cölfen   1

Bio-inspired Crystal Growth by Synthetic Templates
S.-H. Yu 79

Delayed Action of Synthetic Polymers
for Controlled Mineralization of Calcium Carbonate
K. Naka 119

Inorganic–Organic Calcium Carbonate Composite
of Synthetic Polymer Ligands with an Intramolecular NH···O Hydrogen Bond
N. Ueyama · K. Takahashi · A. Onoda · T. Okamura · H. Yamamoto . . . 155

Author Index Volumes 251–271 195

Subject Index 205

Contents of Volume 270

Biomineralization I

Volume Editor: Kensuke Naka
ISBN: 978-3-540-46379-5

Crystallization of Calcium Carbonate Beneath Insoluble Monolayers:
Suitable Models of Mineral–Matrix Interactions in Biomineralization?
M. Fricke · D. Volkmer

Self-Organized Formation of Hierarchical Structures
H. Imai

Fluorapatite-Gelatine-Nanocomposites:
Self-Organized Morphogenesis, Real Structure and Relations
to Natural Hard Materials
R. Kniep · P. Simon

Inorganic-Organic Interfacial Interactions
in Hydroxyapatite Mineralization Processes
K. Sato

Detoxification Biominerals
C. K. Carney · S. R. Harry · S. L. Sewell · D. W. Wright

Bio-inspired Mineralization Using Hydrophilic Polymers

Helmut Cölfen

Department of Colloid Chemistry, Max Planck Institute of Colloids and Interfaces,
MPI Research Campus Golm, 14424 Potsdam, Germany
coelfen@mpikg-golm.mpg.de

1	Introduction	3
2	Different Crystallization Modes and Ways to Modify Crystallization	5
2.1	Classical Crystallization	6
2.1.1	Thermodynamic and Kinetic Crystallization Pathways	7
2.1.2	Face-Selective Additive Adsorption	10
2.2	Non-classical Crystallization	11
2.2.1	Oriented Attachment	12
2.2.2	Mesocrystals	14
2.2.3	Amorphous Precursors	18
2.2.4	Liquid Precursors	20
3	Polymer-Controlled Crystallization	22
3.1	Biomineralization – Some Typical Hydrophilic Polymers	23
3.2	Bio-inspired Mineralization	25
3.2.1	Biopolymers	26
3.2.2	Homopolymers and Random Copolymers	29
3.2.3	Dendrimers	38
3.2.4	Double Hydrophilic Block Copolymers (DHBCs)	39
3.2.5	Double Hydrophilic Graft Copolymers (DHGCs)	64
4	Conclusion	64
5	Current Trends and Outlook to the Future	66
	References	68

Abstract Biomineralization processes result in organic/inorganic hybrid materials with complex shape, hierarchical organization, and superior materials properties. Chemistry, which is inspired by these processes, aims to mimic biomineralization principles and to transfer them to the general control of crystallization processes using an environmentally benign route. In this chapter, the latest advances in hydrophilic polymer-controlled morphosynthesis and bio-inspired mineralization of crystals are summarized with focus on the various principles that can be used to generate inorganic and organic crystals with unusual structural specialty and complexity. For this, classical crystallization pathways using crystal face-selective polymer adsorption can be applied as well as non-classical nanoparticle-mediated crystallization routes, which are based on amorphous or crystalline precursor particles. Current developments emphasize that probably all inorganic and organic crystals will be amenable to morphosynthetic control by the described strategies using either flexible polymer additives or suitable self-assembly mechanisms. The resulting unique hierarchical materials with structural specialty and complexity, and a size range spanning from nanometers to micrometers, are expected to find potential applications in various fields. In addition, bio-inspired mineralization with hydrophilic copolymers offers the chance to understand basic principles of the complex and synergetic biomineralization processes.

Keywords Bio-inspired mineralization · Double hydrophilic block copolymers · Mesocrystals · Non-classical crystallization · Oriented attachment · Self-assembly

Abbreviations
Amino acid abbreviations follow the standard three-letter code

ACC	Amorphous calcium carbonate
AUC	Analytical ultracentrifugation
CMC	Critical micelle concentration
COD	Calcium oxalate dihydrate
CTAB	Cetyltrimethylammoniumbromide
DHBC	Double hydrophilic block copolymer
DLS	Dynamic light scattering
EDTA	Ethylene diamine tetra acetic acid
HAP	Hydroxyapatite
OCP	Octacalcium phosphate
PAA	Poly(acrylic acid)
PAH	Poly(allylamine hydrochloride)
PAM	Poly(acrylamide)
PASM	Poly(ammonium 2-sulfatoethyl methacrylate)
PDADMAC	Poly(diallyldimethylammonium chloride)
PDEAEMA	Poly(diethylaminoethylmethacrylate)
PEI	Poly(ethylene imine)
PEIPA	Poly(ethylene imine)-poly(acetic acid) or poly(EDTA)
PEO	Poly(ethylene oxide)
PEG	Poly(ethylene glycol) or low molar mass PEO
PGL	Poly(glycidol)
PHEA	Polyhydroxyethylacrylate
PHEE	Poly(hydroxyethylethylene)
PHEMA	Poly(2-hydroxyethyl methacrylate)
PHMA	Poly(dihydroxypropyl methacrylate)
PILP	Polymer induced liquid precursor
PMAA	Poly(methacrylic acid)
PMVE	Poly(methylvinylether)
PNIPAM	Poly-(N-isopropylacrylamide)
PPI	Poly(propylene imine)
PSMA	Poly(styrene-alt-maleic acid)
PSS	Poly(sodium 4-styrenesulfonate)
PVA	Poly(vinyl alcohol)
PVOBA	Poly(vinyloxy-4-butyric acid)
PVP	Poly(vinyl pyrrolidone)
SAM	Self assembled monolayer
SANS	Small angle neutron scattering
SAXS	Small angle X-ray scattering

SDS   Sodium dodecylsulfate
SEM   Scanning electron microscopy
SFM   Scanning force microscopy
TEM   Transmission electron microscopy
WAXS   Wide angle X-ray scattering

1
Introduction

Synthesis of inorganic crystals or hybrid inorganic/organic materials with specific size, shape, orientation, organization, complex form, and hierarchy has been a focus of recent interest [1–13] because of the importance and the potential to design new materials and devices in various fields such as catalysis, medicine, electronics, ceramics, pigments, and cosmetics. For example, shape control and structural specialty (size, shape, phase, dimensionality, assembly etc.) have significant relevance for the properties of semiconductor nanocrystals [8], metal nanocrystals [9–12], and other inorganic materials [5–7], which may add additional variables for tailoring the properties of nanomaterials [14]. Much effort has been made towards fabricating one-dimensional nanowires or nanorods [15–21] by using hard templates such as carbon nanotubes [15, 16] and porous aluminum templates [17–19] or by laser-assisted catalytic growth (LCG) [20, 21] as well as controlled solution growth at room temperature or elevated temperature [8, 9, 22–25].

As an alternative strategy, bio-inspired morphosynthesis emerged as an important and environmentally friendly route to generate inorganic materials with controlled morphologies by using self-assembled organic superstructures, inorganic or organic additives, and/or templates with complex functionalization patterns [5–7, 26]. The biospecific interaction and coupling of biomolecules with inorganic nanosized building blocks has shown the potential assembly capabilities for generating new organized materials with strong application impact [27]. The directed self-assembly of inorganic and inorganic/organic hybrid nanostructures has also been emerging as an active area of recent research. It therefore makes a lot of sense to learn more about the natural high technology mineral archetypes and to investigate and mimic the mechanisms of biomineralization.

Biomineralization is the process by which living organisms secrete inorganic minerals in the form of skeletons, shells, teeth etc. It is a rather old process in the development of life and was adapted by living beings probably at the end of the Precambrian, more than 500 million years ago [28]. Biominerals are highly optimized materials with remarkable properties and have attracted a lot of attention recently, not only among materials chemists, because they are considered natural archetypes for future materials. In nature there exists an abundance of biominerals, which are usually composites of inorganic and organic materials in a highly organized form with fascinating shapes and structures [29–33]. For example, pearls, oyster shells, corals, ivory, sea urchin spines, magnetic crystals in bacteria, algae exoskeletons, eggshells, bones, and teeth are typical biological minerals created by living organisms.

During the past decade, exploration of novel bio-inspired strategies for self-assembling or surface-assembling molecules or colloids to generate materials with controlled morphologies with structural specialty, complexity, and related unique properties is among the hottest current research subject in the area of materials chemistry and its cutting-edge fields [26, 34–40]. Biomineralization and bio-inspired morphosynthesis have been rapidly developed into one of the central objectives of biomimetic chemistry [41–45]. It has been the object of intense recent research activity for materials scientists and chemists to learn how to create superstructures resembling naturally existing biominerals with their unusual shapes and complexity [1, 46–48]. Recent reviews report the latest research advances in biomineralization of unicellular organisms with complex mineral structures from the viewpoint of molecular biology [49] or highlight how the interplay between aggregation and crystallization can give rise to mesoscale self-assembly and cooperative transformation and reorganization of hybrid inorganic/organic building blocks to produce single-crystal mosaics, nanoparticle arrays, and emergent nanostructures with complex form and hierarchy [40]. All these advances indicate a large potential for bio-inspired mineralization. It is therefore not surprising that chemists have taken up the challenge to learn from the design of nature's sculptures and try to extend their synthetic toolbox with principles from biomineralization.

In biomineralization processes, one must distinguish between a structural insoluble matrix, which provides a scaffold for the subsequent mineralization process, and a functional soluble matrix. Both can control the crystallization process [50]. The latter is often composed of soluble and charged macromolecules so that the application of hydrophilic polymers and polyelectrolytes is advantageous for achieving control over crystallization reactions.

Use of water-soluble polymers as crystal modifiers for controlled crystallization is widely expanding and is a versatile route for controlling and designing the architectures of inorganic materials. These additives do not act as supramolecular templates like the surfactant type or other rigid templates, which predefine the later hybrid material structure but usually act as a soluble species at various hierarchy levels of the forming mineral hybrid (soluble matrix) [40]. In that way, the application of soluble macromolecules as crystallization additives is bio-inspired and also opens up the possibility to design model systems for understanding parts of the complex biomineralization processes.

This review is organized in a way that first, crystallization itself is discussed, as several non-classical and particle-based crystallization pathways have recently been discovered. This will enable the reader to understand the basic principles of crystallization control in biomineralization and bio-inspired mineralization. As a second part, the possibilities for controlling crystal morphogenesis by hydrophilic polymers are discussed. Conclusions are drawn in the fourth part and the chapter finishes with the identification of current trends and an outlook to the future (Heading 5).

2
Different Crystallization Modes and Ways to Modify Crystallization

Driven by the intensified research on biomineralization processes as inspiration for the synthesis of future advanced materials [30, 32, 51–56] and also bio-inspired mineralization [41, 57], several indications were found that crystallization does not necessarily proceed along the classical textbook crystallization pathway, which is the attachment of ions/molecules to a primary particle forming a single crystal.

Instead, crystallization can also proceed along particle-based reaction channels [40]. Particle-mediated crystallization pathways have been identified that form crystals in the process of a so-called mesoscopic transformation including self-assembly or transformation of metastable or amorphous precursor particles [40]. Such crystallization pathways especially apply to systems far from equilibrium, for which the classical thermodynamic considerations are no longer valid for predicting the morphology or size of the crystals (see also Sect. 2.1.1). But, even for systems considered to crystallize via the classical pathway, indications were found that nanoparticles are involved in the crystallization process. This nanoparticle-mediated crystallization pathway involving mesoscopic transformation is called "non-classical crystallization" in the following text and involves multiple nucleation events of nanoparticles to form a nanoparticle superstructure, in contrast to a single nucleation event to form a single crystal.

Non-classical crystallization often involves self-assembly of preformed nanoparticles to an ordered superstructure, which then can fuse to a single crystal. These pathways are outlined in Fig. 1 and discussed below in more detail.

From Fig. 1, it becomes clear that alternative particle-mediated crystallization pathways can also finally lead to a single crystal (see Oriented Attachment Sect. 2.2.1 and Mesocrystals Sect. 2.2.2 where the nanoparticle units can fuse together by oriented attachment to form a single crystal) Amorphous precursor particles (Sect. 2.2.3), which can crystallize and then form a single crystal via a mesocrystal or directly by oriented attachment and even liquid precursors (Sect. 2.2.4), also belong to these non-classical crystallization pathways. These new and developing concepts are not only expanding our views on crystallization in general, they are also possibly tools for un-

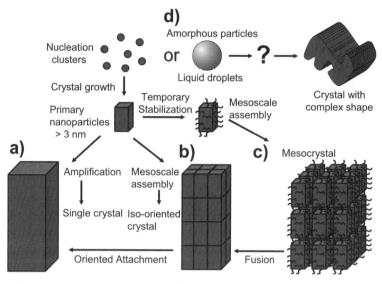

Fig. 1 Schematic representation of classical and non-classical crystallization. *Pathway a* represents the classical crystallization pathway where nucleation clusters form and grow until they reach the size of the critical crystal nucleus growing to a primary nanoparticle, which is amplified to a single crystal (*path a*). The primary nanoparticles can also arrange to form an iso-oriented crystal, where the nanocrystalline building units can crystallographically lock in and fuse to form a single crystal (oriented attachment, *path b*). If the primary nanoparticles become covered by a polymer or other additive before they undergo a mesoscale assembly, they can form a mesocrystal (*path c*). Note: Mesocrystals can even form from pure nanoparticles. There is also the possibility that amorphous particles are formed, which can transform before or after their assembly to complicated morphologies (symbolized by the question mark in *path d*). The scheme is partly based on that in [40]

derstanding biomineralization processes in more detail as they extend the toolbox of crystallization, which has so far been set by the classical crystallization pathway.

2.1
Classical Crystallization

The discussion of non-classical crystallization processes requires a brief consideration of classical crystallization itself at the very basic level. Crystallization starts from ions in the case of inorganic minerals or molecules. These primary building units form clusters, which can grow or disintegrate again because of the counter-play between the surface energy ($\sim r^2$) of the newly created surface and the bulk energy gained from the formation of a crystal lattice ($\sim r^3$). The clusters eventually reach the size of a so-called critical crystal nucleus, at which the surface energy is counterbalanced by the bulk energy so that the change in the free enthalpy of the system becomes negative upon further particle growth as the win in lattice energy overcompensates the loss in surface energy. The critical crystal nucleus is the smallest crystalline unit capable of further growth. The so-formed primary nanoparticles grow further by ion attachment and unit cell replication (Fig. 1 pathway a).

The shape of inorganic crystals is normally related to the intrinsic unit cell structure and the crystal shape is often the outside embodiment of the unit-cell replication and amplification. The diverse crystal morphologies of the same mineral are due to the differences in the crystal faces with regard to surface energy and external growth environment, as already revealed early last century by Wulff [58]. Generally speaking, the growth rate of a crystal face is usually related to its surface energy if the same growth mechanism acts on each face. The fast growing faces have high surface energies and they will vanish in the final morphology, and vice versa. This treatment assumes that the equilibrium morphology of a crystal is defined by the minimum energy resulting from the sum of the products of surface energy and surface area of all exposed faces (Wulff rule). Although it is nowadays well known that this purely thermodynamic equilibrium treatment cannot always predict the crystal morphology, as crystallization and the resulting morphology often also relies on kinetic effects as well as on defect structures like screw dislocations or kinks etc., it forms the basis for the consideration of additive-mediated crystal morphology changes (see also Sect. 2.1.2).

2.1.1
Thermodynamic and Kinetic Crystallization Pathways

As mentioned above, consideration of the thermodynamic equilibrium is not very well suited to describe the crystallization pathways observed in biomineralization and biomimetic mineralization processes. Instead, it can be stated that these systems are usually far from equilibrium so that kinetic effects prevail over thermodynamic control, with the consequence that it is still almost impossible to predict the morphology of a crystal with the currently existing thermodynamic equilibrium models, which are more or less a reflection of Wulff's rule [58].

In general, kinetic control is based predominantly on the modification of the activation-energy barriers of nucleation, growth, and phase transformation (Fig. 2) [32]. In such cases, crystallization often proceeds by a sequential process involving structural and compositional modifications of amorphous precursors and crystalline intermediates, rather than a single-step pathway [40, 56, 59, 60]. How far these metastable intermediates become stabilized by the additive or are transformed to the next stable species according to Ostwald's step rule depends on the solubility of the minerals and on the free energies of activation of their interconversions, all of which are strongly influenced by additives [35]. The corresponding changes in composition and

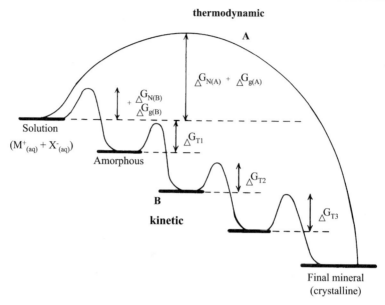

Fig. 2 Crystallization pathways under thermodynamic and kinetic control. Whether a system follows a one-step route to the final mineral phase (*pathway A*) or proceeds by sequential precipitation (*pathway B*), depends on the free energy of activation associated with nucleation (N), growth (g), and phase transformation (T). Amorphous phases are common under kinetic conditions. Reproduced from [40] with permission of Wiley

structure usually occur by dissolution–renucleation processes closely associated with the surface and/or the interior of preformed particles, as demonstrated for calcite hollow spheres grown on the surface of sacrificial vaterite spheres [61]. The nucleation of a particular crystalline phase is therefore highly heterogeneous and dependent on interfacial and hydrodynamic properties as well as on reaction kinetics. Therefore, this multistep process cannot be readily described by the classical model of nucleation, which is based on a thermodynamic treatment of the interplay between energy gain through bulk crystallization and energy loss through increase in surface area, namely the classical Wulff treatment [58].

Kinetically driven crystallization often involves an initial amorphous phase that may be non-stoichiometric, hydrated, and susceptible to rapid phase transformation. Amorphous calcium carbonate (ACC) for instance is highly soluble, has a low density of almost half of the crystalline mineral indicating a high hydration [62], and rapidly transforms to calcite, vaterite, or aragonite unless kinetically stabilized. In aqueous solution, this transformation into vaterite or calcite takes place within seconds or less even if additives are present, as shown by recent SAXS/WAXS measurements of ACC transformation in the presence of a DHBC [63].

In biomineralization, the kinetic ACC stabilization is thought to be achieved by ions such as Mg^{2+} and PO_4^{3-}, or by enclosing the amorphous phase in an impermeable sheath of organic macromolecules, such as polysaccharides [64] or mixtures of polysaccharides and proteins rich in glutamic acid, threonine, and serine [65, 66]. Thus, a significant number of stabilized ACC biominerals have recently been documented in plant cystoliths [67, 68], snail shells [69], aragonitic mollusk shells [70, 71], nacre [72], ascidian spicules [51, 65, 73], precursors of the carbonated apatite inner tooth layer of chitons [74], and crustacean exoskeletons [75] and likely, more biominerals belong to this list.

Kinetic control of crystallization can be achieved by modifying the interactions of nuclei and developing crystals with solid surfaces and soluble molecules [76]. Such processes influence the structure and composition of the nuclei, particle size, texture, habit and aggregation, and stability of intermediate phases. In biomineralization, for example, structured organic surfaces are considered to play a key role in organic matrix-mediated deposition by lowering the activation energy of nucleation of specific crystal faces and polymorphs through interfacial recognition [77, 78]. Soluble macromolecules and organic anions, as well as inorganic ions such as Mg^{2+} and PO_4^{3-}, can also have a marked kinetic effect on crystallization, particularly with regard to polymorph selectivity and habit modification. For example, magnesium or polyphosphonate can inhibit $CaCO_3$ crystallization resulting in ACC [79–84]. These interactions can be highly specific; for example, proteins extracted either from calcite or aragonite-containing layers of the abalone shell induce the crystallization of calcite or aragonite, respectively, when added to supersaturated solutions of calcium hydrogen carbonate in the laboratory with or without being immobilized on a surface [50, 85]. Atomic-force microscope (AFM) studies [86] indicate that the macromolecules bind to growth sites on well-defined crystal surfaces and influence the kinetics of crystal growth from solution in accordance with the classical model of secondary nucleation [87]. These studies on bulk crystalline system growth indicate that polymer additives act in a classical, and thus predictable, way as selective surface poisoning agents. This is further supported by the results of functionalized latex adsorption onto growing crystals [88, 89] so that they can even become occluded into a single crystal, which can become an interesting porous material after latex removal [90, 91]. However, it is not yet clear if these results can be transferred to the more relevant case of crystalline nanoparticles at the early experimental stage, shortly after nucleation, which are the relevant species for the outcome of a polymer-directed crystallization reaction for many biomineralization and biomimetic mineralization processes.

2.1.2
Face-Selective Additive Adsorption

The thermodynamic consideration of crystallization discusses the crystal morphology as a result of energy minimization of all exposed faces (Wulff rule) [58]. The surface energy of a crystal face can be lowered by the adsorption of an additive. Therefore, the shape of crystals can be affected by various factors, i.e., inorganic ions or organic additives. The anisotropic growth of the particles can be explained by the specific adsorption of ions or organic additives to particular faces, therefore inhibiting the growth of these faces by lowering their surface energy.

This strategy of crystal morphogenesis has been known for a long time and has even found industrial application, mainly based on empirical observations. The striking influence of small molecule additives on the shape and structures of inorganic crystals has been well documented. These small additives include simple inorganic ions, small molecular organic species, and solvents, which have been found to exert strong influence on the shape of inorganic crystals. However, the additives can also be much more complex and even be designed to bind to specific crystal faces.

Laursen and DeOliveira have shown an excellent example of using protein secondary structures to control the orientation of chemical functionality so that the protein can bind to a targeted crystal face. An α-helical peptide (CBPI) with an array of aspartyl residues was designed to bind to the ($1\bar{1}0$) prism faces of calcite [92]. The careful observation of the effect of CBP1 and other peptides on calcite crystal growth was skillfully done by adding the peptide to rhombohedral seed crystals growing from a saturated $Ca(HCO_3)_2$ solution. When CBP1 was added to seed crystals (as shown in Fig. 3a) and growth was allowed to continue, the calcite crystals elongated along the [001] direction (c-axis) with rhombohedral (104) caps (Fig. 3b). After washing the crystals with water and replacing the mother solution with fresh saturated $Ca(HCO_3)_2$ solution, a regular rhombohedron formed by a sort of repair process with subsequent growth on the putative prism surfaces (Fig. 3c). CBP1 was only about 40% helical when the temperature was 25 °C, leading to the formation of studded crystals by epitaxial growth perpendicular to each of the six rhombohedral surfaces (Fig. 3d and e). After washing these crystals and regrowing in fresh $Ca(HCO_3)_2$ solution, repair of the non-rhombohedral surfaces was again observed. In each study, a new rhombohedron was formed and thus six regular rhombohedra overgrew the original seed (Fig. 3f).

If additive adsorption occurs very selectively, even one-dimensional nanostructures can be grown, as demonstrated for Ag nanowires with PVP adsorbing onto the side faces of the elongated nanoparticle. Thus the surface energy of these faces was lowered so that the Ag rods only grew at their ends [93]. However, if the polymer adsorption is not homogeneous and has defects, branching may occur as shown later in Fig. 5a.

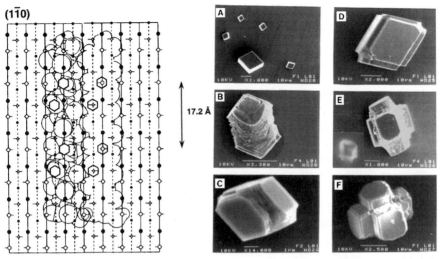

Fig. 3 *Left*: The footprint of two α-helical peptide (CBP1) molecules binding to the (1̄10) prism faces of calcite. The *filled circles* are Ca^{2+} ions, and *open circles* are CO_3^{2-} ions. *Large circles* are ions in the plane of the surface and *small circles* are 1.28 Å behind this plane. The *hexagons* indicate that peptide carboxylate ions occupy CO_3^{2-} sites on the corrugated surface. *Right*: SEM micrographs showing the effect of CBP1 on the growth of calcite crystals: **a** Calcite seed crystals showing typical rhombohedral morphology. **b** Elongated calcite crystals formed from seed crystals at 3 °C in saturated $Ca(HCO_3)_2$ containing ca. 0.2 mM CBP1, showing expression of new surfaces corresponding to the prism faces and rhombohedral (104) caps; growth is along the c-axis, which extends through the vertices of the end caps. **c** "Repair" and re-expression of rhombohedral surfaces when crystals from **b** are allowed to grow in saturated $Ca(HCO_3)_2$ after removal of CBP1 solution. **d,e** Respective earlier and later stages of growth of calcite crystals from rhombohedral seed crystals at 25 °C in saturated $Ca(HCO_3)_2$ containing ca. 0.2 mM CBP1, showing epitaxial growth from the rhombohedral surfaces. **e** *inset*: Fluorescence micrograph of epitaxial studded calcite crystals grown under similar conditions in the presence of fluorescein-labeled CBP1; the *dark spot* in the center corresponds to a (104) face, indicating that the peptide is binding to the newly expressed surfaces and not (104). **f** "Repair" and re-expression of rhombohedral surfaces when crystals from **e** are allowed to grow in saturated $Ca(HCO_3)_2$ after removal of CBP1. Reproduced in part from [92] with permission of the American Chemical Society

2.2
Non-classical Crystallization

The particle-mediated non-classical crystallization path makes crystallization more independent of ion products or molecular solubility. It can occur without pH or osmotic pressure changes, and opens new strategies for crystal morphogenesis. This is possible because the precursor particles can be formed independently and can be transported to the mineralization site so that, at the site of crystallization, no supersaturation with its associated elevated ion concentrations is required and pH changes coupled to a crystallization reaction can be avoided. As a result of these features, nanoparticle-mediated non-classical crystallization should be especially relevant in the process of biomineralization in biological systems because of the obvious advantages for a living system, which has to handle the control of a crystallization reaction. However, the identification of non-classical crystallization processes in biomineralization processes is very difficult as many biominerals are not static structures but are continuously remodeled according to time-dependent demands, as for example in bone. In addition, the precursor structures of the particle-mediated non-classical crystallization pathways are difficult to detect as they are often only of a transient nature and can fuse together to form a single crystal by the process of oriented attachment (see also Sect. 2.2.1) involving crystallographic fusion of mutually perfectly aligned nanocrystals to a single crystal. Nevertheless, in bio-inspired mineralization, non-classical crystallization is easier to study and thus better documented than for biomineralization. However, the study of non-classical crystallization is still in its infancy as these crystallization routes were only discovered a few years ago.

2.2.1
Oriented Attachment

Oriented attachment was first found for TiO_2 particles generated in a hydrothermal process [94]. Chains of nanoparticles fused in crystallographic register were found and the fusion surfaces were identified to be the highest energy surfaces of the nanoparticles (Fig. 4).

From the thermodynamic viewpoint, the system wins a substantial amount of energy by eliminating two high energy surfaces by crystallographic fusion, which is the driving force for the directed particle fusion [95, 96]. This can happen if two nanoparticles approach each other so closely that they mutually attract by van der Waals attraction. However, they can still have enough thermal energy to wiggle around next to each other until, eventually, the high

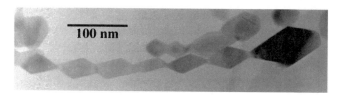

Fig. 4 TEM micrograph of a single crystal of anatase that was hydrothermally coarsened in 0.001 M HCl, showing that the primary particles align, dock, and fuse to form oriented chain structures. Reproduced from [94] with permission of Elsevier

energy surfaces come into direct contact, lock in, and crystallographically fuse together eliminating the two high energy surfaces.

In the meantime, oriented attachment was found for a variety of systems such as TiO_2 [94, 97–99], FeOOH [100], CoOOH [101], Ag [102], ZnO [103], and ZnS [104] and was recently summarized in a review paper [105].

Oriented attachment is not only described for the one-dimensional case [94, 98, 99, 103], but is also described for two- [106] and three-dimensional cases [107]. Oriented attachment offers the special advantage of producing defect-free one-dimensional single crystals, which is certainly of interest in materials science. Classical crystallization always engenders defect structures in the form of branches when attempting fiber growth by additive adsorption to all crystal faces parallel to the growth axis and thus these faces become blocked from further growth. In contrast, oriented attachment offers the crystallographic fusion of nanoparticles to single crystalline and defect-free fibers, which can be hundreds of micrometers long. This is demonstrated in Fig. 5. Figure 5a shows the example of hydroxyapatite (HAP) fibers, which grow in an aggregate of special block copolymers, adsorbing to all faces parallel to the c-growth axis [108].

Although thin and long crystalline HAP fibers were obtained (the single crystalline nature of the fibers could not be shown due to their small diam-

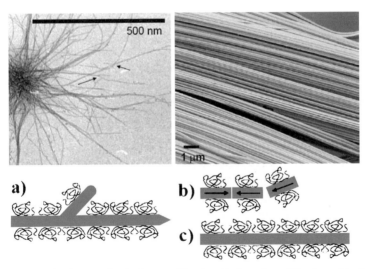

Fig. 5 Classical (**a**) vs. non-classical crystallization (**b,c**). **a** Crystallization of hydroxyapatite (HAP) fibers from block copolymer aggregates [108], where the block copolymers adsorb to all faces parallel to the HAP c-growth axis resulting in whisker structures with occasional branches (see *arrows*). **b,c** Formation of single crystalline and defect-free $BaSO_4$ (210) oriented fiber bundles by the process of oriented attachment in experiments described in [109–111]. Figure reproduced from [112] with kind permission of Editorial Universitaria, Universidad de Chile, Santiago, Chile

eter of only 2–3 nm), the fibers clearly show occasional branches. These can be attributed to a non-continuous polymer layer, so that branching can occur at the non-covered sites on the crystal as result of the ion-based classical growth mechanism. If on the other hand, all crystal faces parallel to one axis of a nanoparticle are already covered by polymers, only two opposite faces remain uncovered and thus become high energy faces. These faces can then fuse together to form a single crystalline fiber without any defects following the mechanism of oriented attachment (Fig. 5b,c). Here, the advantages of the particle-based mechanism compared to the ion-based one become directly obvious.

Although, so far, there are still comparatively few reports on the oriented attachment of nanoparticles, this new growth mechanism offers an additional tool for the design of single crystalline advanced materials with anisotropic shape and it is certainly also worth exploring how far oriented attachment plays a role in biomineralization.

2.2.2
Mesocrystals

Mesocrystals are colloidal crystals and are built up from individual nanocrystals, which are aligned in a common crystallographic register (see also Fig. 1 pathway c) so that a mesocrystal scatters like a single crystal [113]. Probably, mesocrystals are much more common than assumed so far but it is difficult to detect them as they may be misinterpreted as single crystals due to their single crystal scattering properties and their well-facetted appearance [113] (see also Fig. 6 for an example). Additionally, if the surface of the nanocrystals is not sufficiently stabilized, a mesocrystal can easily transform to a single crystal by oriented attachment (Fig. 1 and Sect. 2.2.1), as the nanoparticle units are already crystallographically aligned so that a crystallographic fusion is thermodynamically favored.

Mesocrystals are a very interesting form of colloidal crystal, as they extend the so far known colloidal crystals with spherical building units to those with non-spherical building units. This offers new possibilities of superstructure formation due to the anisotropic particle shape of the nanoparticle building units [113]. Thus, mesocrystals are colloidal crystals but with extended possibilities for their self-assembled superstructure, offering new handles for crystal morphology control.

The mesocrystal concept has just very recently been developed for minerals grown by a bio-inspired mineralization process (as reviewed in [113]) but it also appears to be of importance in biomineralization processes, as very recently proposed for nacre with aragonite tablets built up from mutually perfectly crystallographically aligned 45 nm nanograins [114]. In addition, the construction of a natural nacre aragonite platelet from mutually oriented nanoparticles was also reported by Oaki and Imai [115]. A further example is

Fig. 6 Fluoroapatite grown in a gelatin gel. **a** SEM of the hexagonal prismatic seed together with the corresponding diffraction pattern. The *arrow* indicates the direction of the incident X-ray beam. **b** Gelatinous residue of the composite. **c** SEM image of the fracture area of a central seed. **d** Arrangement of hexagonal nanoparticles forming a superstructure. The *lines* and *arrows* indicate preferred cleaving directions. Figures **a,b,d** are reproduced from [119] with permission of Wiley. Figure **c** is reprinted from [120] with permission of the Royal Society of Chemistry

aligned nanobricks in sea urchin spines, which could clarify the long debate on why the spines diffract like a single crystal but show the fracture behavior of a glass [116].

One of the most investigated synthetic mesocrystals is the hexagonal prismatic seed crystal of fluoroapatite, formed in a gelatin gel, which further grows to spherical particles via dumbbell intermediates (Fig. 6a,b) [117, 118]. The hexagonal seed shows all the typical features of a mesocrystal and is thus a nice example for demonstrating the basic properties of a mesocrystal and also the problems of identification (Fig. 6a). The hexagonal seed crystal was not directly recognizable as a mesocrystal, as it showed a well-facetted, single crystal-like morphology (Fig. 6a). Even X-ray diffraction showed features of a fluorapatite single crystal oriented along the c-axis [119], due to the very high vectorial order of its nanoparticulate building units (Fig. 6a). After gelatin crosslinking and apatite removal, the polymer distribution inside the crystal could be visualized replicating the structure of the former dumbbell-shaped hybrid particle (Fig. 6b). The internal structure of the hexagonal seed crystals (Fig. 6c) revealed a radial pattern and a superstructure periodicity of 10 nm, in good agreement with a primary nanoparticle size of about 10 nm [120]. In addition, structural defects attributed to a collagen triple helix strand were detected, as well as self-similar nano-subunits nucleated by gelatin [121]. On basis of this evidence, the growth model for the observed radial outgrowth in the hexagonal seed was developed (Fig. 6d). It agrees with the mesocrystal scheme presented in Fig. 1 pathway c, but for hexagonal building units.

Although the above system has been very well investigated, the formation process of the mesocrystal is still not fully revealed. This is essentially true for most reported mesocrystals. Although precursor nanoparticles have already been found experimentally [122] and dipole fields are suggested to be responsible for their almost perfect alignment [117, 123], the full growth mechanism has so far only been reported for two examples.

One example is a copper oxalate mesocrystal [124]. Here, nanoparticles were found to arrange almost perfectly to a mesocrystal, which could be influenced in terms of morphology by hydroxymethylpropylcellulose (HPMC). The polymer influences nucleation, nanocrystal growth and aggregation by face-selective interaction. Upon aggregation of the nanocrystals, a mesocrystal is formed as intermediate but is apparently not stable due to the low repulsive electrostatic and steric forces. Depletion flocculation of the weakly adsorbed polymer layers was suggested to be the reason for the depletion of the polymer from the inner mesocrystal surfaces to the outer mesocrystal faces, resulting in attraction of the nanoparticles with subsequent nanoparticle fusion towards an iso-oriented crystal. As typical for mesocrystals, electron diffraction indicated a minor, but detectable, orientational disorder that also supported the nanoparticle self-assembly based mechanism for mesocrystal formation. Later, the "brick-by-brick" self-assembly mechanism was experimentally revealed in a time-dependent study [125]. The whole mesocrystal formation process is visualized in Fig. 7.

Mesocrystals can even be formed without additive, as found for the related cobalt oxalate dihydrate [126]. Copper oxalate crystals were composed of nanometer building units and strings of nanodomains oriented along the principal axis of the particle were detected, revealing that the lateral and basal faces of the precipitate are composed of stacked nanoparticle layers with a thickness of 5–7 nm.

A time-resolved HRSEM study revealed the formation mechanism of the mesocrystal [126]. Poorly crystalline primary particles (10 nm) first aggregate to form secondary particles (23 nm). The latter subsequently aggregate to form polydisperse elongated particles. These elongated particles also ag-

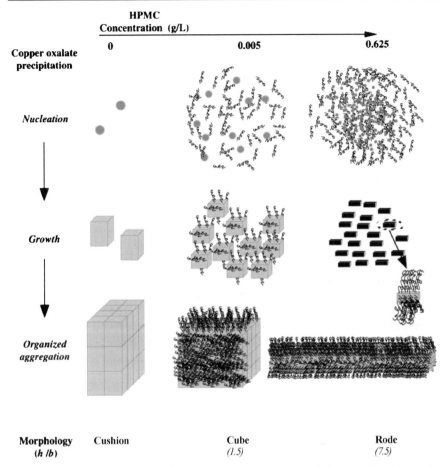

Fig. 7 Schematic representation of copper oxalate precipitation. Influence of HPMC on the three major steps of particle formation. Figure reproduced from [124] with permission of Academic Press

gregate at the end and center of a growing particle, building the mesocrystal. Afterwards, there is a layer-by-layer growth by aggregation of primary particles onto the lateral external faces. Thus, the mesocrystal is a core–shell particle or core–shell mesocrystal, where the core is quite disordered due to the polydispersity of the nanoparticle building units [126]. The presence of steps and kinks on the external faces is reminiscent of the classical crystallization models – except that the atoms or molecules are replaced by nanoparticles. The evolution of supersaturation and thus ionic strength, closely associated to colloidal stability of the otherwise unstabilized nanoparticles, were suggested to play a predominant role in the ordering process of the nanocrystals to a mesocrystal as the ordered mesocrystal shells were formed at a lower supersaturation than the disordered cores.

Mesocrystals can in principle not only be formed as a polymer–nanocrystal two phase system but also from any other two phase system, where one phase has to be the mutually oriented crystalline nanoparticle phase. The other phase can be any other type of solid phase – the most important and relevant case in view of biomineralization being an amorphous phase in between the aligned nanoparticle crystals, although such phase has not yet been proven for mesocrystals. That such thin amorphous layers can really exist was recently shown by a HRTEM study of synthetic aragonite [127] and of *Haliotis Laevigata* gastropod nacre [72]. It was suspected that this layer was formed by accumulation of impurities, which prevented further crystallization in analogy to the zone melting process in metallurgy. A composite material between aligned single crystalline nanoparticles embedded in an amorphous layer containing the impurities and including polymers therefore seems to be an intriguing hypothesis for biominerals such as sea urchin spines, which scatter like single crystals (the aligned nanocrystal phase) but fracture like an amorphous glass (the interparticle phase). Up to now, the glass-like fracture behavior was explained as a result of the protein inclusions [128] and it remains to be investigated how far such systems are mesocrystals. Recent investigations suggested this for skeletal elements of sea urchins as well as for sponge spikes [129, 130] although the authors of these papers suggest a continuous crystalline phase. This indicates the degree of uncertainty that is still related to the investigation of mesocrystals and the particular analytical problems associated with the analysis of such systems. Further complications arise from the transient nature of mesocrystals.

A mesocrystal is usually only an intermediate in the formation pathway of a single crystal. Fusion of the oriented nanoparticle building units as a result of the high total surface energy of the system will lead to a single crystal by the process of oriented attachment under displacement of the stabilizing polymers, eventually leaving parts of the initially nanoparticle-adsorbed organic components included inside the single crystal.

2.2.3
Amorphous Precursors

Another exciting possibility for creating complex morphologies via a nonclassical crystallization process involving mesoscale transformation of a nanoparticle is the generation of amorphous precursor particles ([40, 54, 56], Fig. 1 pathway d). This strategy is especially capable of generating complex crystal morphologies because amorphous precursors can be molded into any form. The crystallization pathway via amorphous precursor particles is also used in biomineralization [56, 131] and is possibly more common than known to date. Amorphous precursor particles offer the advantage of materials storage in biological systems independently of osmotic stresses generated by high salt concentrations, restrictions of pH or ionic strength constraints, and with the material being available in high concentration at a mineralization site.

In synthetic bio-inspired mineralization systems, amorphous precursor particles are omnipresent [57] so that they are often the early nanoparticle stage in a non-classical crystallization pathway. However, these transient species are difficult to observe in synthetic systems, and it seems very challenging to observe them in a forming biomineral. In this situation, it is promising to learn more about a biomineralization process by applying a bio-inspired model system. As one example, the retrosynthesis of nacre is given as a synthetic model system, where the demineralized insoluble matrix of a biomineral was used for filling with the mineral via amorphous precursors. Nacre from *Haliotis laevigata* was demineralized in acetic acid and subsequently remineralized with amorphous $CaCO_3$ nanoparticles, which were formed in the presence of $\mu g\,mL^{-1}$ amounts of polyaspartic acid [132]. Via these amorphous precursors, we were able to completely fill the compartments of the insoluble organic matrix with $CaCO_3$. Calcite was obtained instead of the natural aragonite and the single crystalline domains were smaller than for the natural archetype (Fig. 8). Nevertheless, the retrosynthesized nacre was indistinguishable from its natural archetype by electron microscopy (SEM and TEM) as shown in Fig. 8. This experiment suggests that (i) natural nacre is built up by amorphous precursor particles, (ii) no specialized protein functions are needed to achieve the generation and transport of

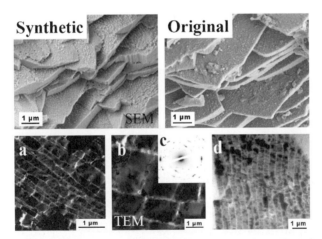

Fig. 8 Nacre retrosythesized in the insoluble organic matrix of a *Haliotis laevigata* shell via amorphous precursor particles. The *upper pictures* compare the natural and synthetic nacre. The *lower pictures* show the corresponding transmission electron micrographs (TEM) of microtomed samples: **a** synthetic nacre, **b** synthetic nacre in higher zoom, **c** electron diffraction pattern taken in this region and showing an almost single crystalline calcite domain, and **d** thin cut of natural nacre. Reprinted from [132] with permission of the American Chemical Society

the amorphous precursor material to the mineralization site – a simple polyelectrolyte function is sufficient, and (iii) the polymorph is controlled by the soluble matrix, which was removed in this set of experiments. Indeed, aragonite could be nucleated in presence of soluble proteins extracted from an aragonite nacre layer [50, 85].

2.2.4
Liquid Precursors

One of the most fascinating mechanisms for generating complex crystalline morphologies is the pathway via liquid precursors. Liquid–liquid phase separation can be observed as an undesirable effect during protein crystallization from solution [133] and has been known for more than a century for $CaCO_3$ [134–136]. If such a phase is used as a precursor phase for a subsequent crystallization process, crystal morphologies with complex shape, molten appearance, and crystalline coatings can be realized.

Such a liquid–liquid phase separation of a saturated crystallization solution can be induced with tiny amounts ($\mu g\,mL^{-1}$ range) of polyelectrolytes like PAsp or PAA in case of $CaCO_3$. Gower et al. proposed the term "polymer-induced liquid-precursor (PILP) process" for the formation process of this type of precursors [137]. They can be observed as little droplets in the crystallization solution by light microscopy. Such liquid precursor droplets were independently observed with anionic dextran sulfate as additive [138]. The role of the polymer appears to be the sequestering and concentration of the calcium ions while simultaneously delaying crystal nucleation and growth to form a metastable solution.

As the resulting crystalline products retain the shape of the precursor, non-equilibrium morphologies can be generated (Fig. 9). The amazing spherulitic vaterite aggregates with helical extensions with spiral pits were suggested to

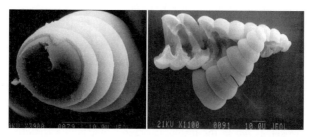

Fig. 9 *Left*: $CaCO_3$ helix formed in the presence of polyaspartate. The helix has been fractured, revealing an outer membrane and polycrystalline core. The helices were fractured using a micromanipulator on an optical microscope. *Scale bar* represents 10 µm. *Right*: Hollow $CaCO_3$ helices fractured by micromanipulation. The fracture is different from that of the solid structures, as shown on the *left side*. *Scale bar* represents 10 µm. Reprinted from [137] with permission of Elsevier

grow inward from the deposited $CaCO_3$ films caused by the polymer. These could act as the membranous substrate for the further growth of $CaCO_3$ crystals as a result of elasticity effects of a gelatinous precursor membrane or of a disclination effect in the partially crystalline membrane. Such interesting phenomenon is quite similar to that occurring in some biogenic minerals such as the otoliths of fishes [139, 140]. Here, spherulitic growth is induced by nucleation sites distributed along a surface, and the aragonite or vaterite crystals grow with their c-axes approximately perpendicular to the surface [139].

Despite the very interesting nature of the PILP process, the reason why liquid droplets form, and the direct proof that they are indeed liquid, remained an open question for the classical $CaCO_3$ system. In contrast, the liquid nature of the droplets became directly accessible for an organic PILP system with glutamic acid after droplet fusion by centrifugation [141]. However, the reason for the PILP droplet formation is still unclear. In a recent paper, Wegner et al. postulate a liquid–liquid phase separation with a lower critical solution temperature (LCST) point at about 10 °C in a saturated $CaCO_3$ solution without additives. They postulated a virtual phase diagram in a coordinate system spanned by temperature and CO_2 concentration [142, 143]. The term virtual means here that the natural consideration of $CaCO_3$ crystallization was not considered in this phase diagram as a reflection of the problem of the thermodynamic description of kinetic processes, as outlined in Fig. 2, and additionally that the postulated liquid phase could not be directly experimentally proven. Nevertheless, the proposed virtual phase diagram means that under isothermal conditions and constant Ca^{2+} concentration, formation of the experimentally observed spherical hydrated ACC particles of the postulated liquid character will begin at the binodal at point B

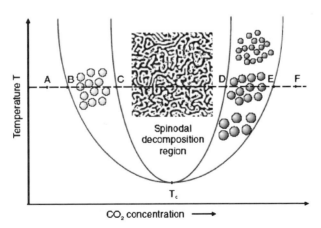

Fig. 10 Schematic virtual phase diagram that explains the formation of spherical particles by liquid–liquid phase segregation. Reproduced from [142] with permission of Wiley

in Fig. 10 up to the spinodal decomposition point C under simultaneous formation of a highly concentrated $CaCO_3$ phase in water. The reverse behavior at still increasing CO_2 concentrations (region D–E) as can be seen in Fig. 10.

Despite all remaining uncertainties about the exact formation mechanism of the liquid mineral precursor droplets, what seems to be clear is that the PILP process enables production of crystalline non-equilibrium morphologies resembling solidified melts or liquid crystals as they form from a fluid-like precursor state. As a result of the shape flexibility of the liquid precursors, the PILP process could be successfully used to produce fairly monodisperse $CaCO_3$-coated oil core–shell particles [144]. This also illustrates that almost all kinds of templates should be suitable for $CaCO_3$ coating via a PILP precursor. $CaCO_3$ mineralization of collagen via a PILP process resulted in fibrous aggregates growing from isotropic gelatinous globules and the fibers became single crystalline with time so that it was suggested that PILP processes may play a role in biomineralization [145].

3
Polymer-Controlled Crystallization

Among the various possible additives for the control of crystallization events, polymers as a class of "surfactants", stabilizers, functional additives or a soft template have been found to play key roles in bio-inspired approaches for mimicking the biomineralization process. The most important role of hydrophilic polymers in that respect is certainly that of a novel surfactant or particle stabilizer. Such surfactants are not always surface-active in a classical sense (i.e., hydrophilic–hydrophobic being attracted by the air/water or oil/water interface), but can also become selectively attracted by mineral or metallic surfaces, only in aqueous solution [146, 147].

In recent years, hydrophilic biopolymers and various synthetic polymers have been widely used for polymer-controlled crystallization and morphosynthesis of different inorganic minerals with still increasing application fields. The principle of controlled crystallization and morphosynthesis in the presence of a soluble polymer not forming a self-assembled superstructure is completely different from the transcriptive template effect provided by pre-designed artificial polymer matrices or templates. The structure formation relies on a synergetic effect of the mutual interactions between functionalities of the polymers and inorganic species, and the subsequent reconstruction in a "programmed" or "coded" way [35, 40, 146].

Biopolymers are often a natural soluble additive choice for morphogenesis of complex crystalline superstructures [148] due to their role in biomineralization, even though it is still difficult to reproduce complex biomineral shapes because an insoluble matrix strongly influences the crystallization location as a compartment. Often, synthetic polymers with precisely known functions and solution behavior are the better choice in terms of a suitable and simplified model system to understand the polymer–mineral interactions. However, their synthesis is often more complex, like that of block copolymers. Therefore, in recent years, biopolymers, homopolymers, and simple random or alternating copolymers have been increasingly applied as crystallization additives with good success after some basic principles have been revealed for block copolymer model systems.

In the following sections, recent advances in the area of bio-inspired mineralization via morphosynthesis of various inorganic minerals by use of biopolymers, various synthetic polymers, and block and graft copolymers will be discussed with a clear focus on recent developments in this rapidly emerging field. Some recent reviews at least partially covering the focus of the present overview exist and shall be mentioned here to allow the reader to find further detail [26, 34–36, 40, 43–45, 57, 105, 113, 149–152].

3.1
Biomineralization – Some Typical Hydrophilic Polymers

The functional biopolymer matrix in biominerals is a multicomponent mixture with the possibility of all kinds of interactions and this makes the analysis of the individual polymer components difficult. Also, the biomineralization proteins are often polydisperse, have a non-globular shape and multiple charges and post translational modifications, which further hampers the protein separation. It is even more difficult to reveal their function as they are not only present in a polymer mixture with the associated possibility of polymer interactions, but are furthermore often only active for a certain time. For example, in calcified parts of crustacean cuticles, 33 different proteins have already been identified [153] so that it is difficult to reveal the functions of individual biomineralization polymers.

Nevertheless, whereas the active proteins for silica biomineralization are cationic to catalyze the silica polycondensation [154], those found to interact with Ca–biominerals are anionic to maintain an interaction with Ca^{2+}.

Important molecules of this class are dentin phosphoryns or sialoproteins containing high aspartic acid (Asp) and extremely high serine (Ser) amounts, which can be phosphorylated by the enzymes casein kinases I and II. Examples include rat dentin phosphoprotein with 43% Ser and 31% Asp [155], rat phosphoryn with 56% Ser and 32% Asp [156], human dentin phosphophoryn with 58% Ser and 26% Asp [157] or a molluscan shell glycoprotein with 32% Ser and 20% Asp [158]. It seems that phosphoserin (P-Ser) is of especial importance for polymer–crystal interactions, very likely in blocked sequences. In addition, the so-called Asprich proteins were recently identified in nacre and can contain up to 53 mol % Asp and 11% Glu [159, 160].

Polysaccharides are of importance as glycoaminoglycans (GAGs), which are polysaccharides bound to a protein core found in vertebrate connective tissues. GAGs are alternating copolymers of a hexosamine and either a galactose or an alduronic acid. Individual GAGs differ from each other by the type of hexosamine or alduronic acid (or hexose), the position and configuration of the glycosidic bonds and the degree and pattern of sulfation. The main acidic building blocks of GAGs are keratin sulfate, hyaluronic acid, chondroin-4-sulfate, chondroin-6-sulfate, dermatan sulfate, heparin sulfate, and heparin [161].

Thus, polysaccharides are usually of importance when they covalently bind to proteins (glycoproteins for example in eggshell mineralization). They only rarely occur unbound as a soluble matrix in biomineralization and often have an alternating structure, usually of the AB copolymer type. They are therefore closer to the structure of polyelectrolytes than to that of proteins/peptides with multiple domain structures and functionalities.

Glycoproteins are also not too well studied as biomineralization polymers although quite a number of biomineral associated proteins are glycosylated [162–165] and evidence is reported that oligosaccharide chains interact with biominerals and/or lectin-like domains of other biomineralization proteins [166, 167]. For example, the calcite binding protein lithostathine consists of a folded C-type lectin core domain and a random coil glycosylated mineral interaction domain consisting of 11 amino acids: pyro-(Glu-Glu-Asp-Gln-Thr-Glu-Leu-Pro-Gln-Ala-Arg) [162]. Glycoproteins are also of importance in antifreeze proteins, which are ice inhibition proteins found in a large variety of organisms where they can show an extended glycosylated helix like in the type IV antifreeze proteins [168–170]. A very interesting glycoprotein is mucoperlin [171], an acidic polymer extracted from *Pinna nobilis* nacre with a short N-terminus, a long set of 13 almost identical tandem repeats enriched in serine and proline residues, a C-terminus with short acidic motifs, three cystein residues, and two potentially sulfated tyrosine residues. The tandem repeat domain offers 27 sites for O-glycosylation at the serine units. Its function could be $CaCO_3$ inhibition and Ca binding. Glycoproteins are also found in the biomineralization process of avian eggshells. For example, on the outer sites of the eggshell membranes, discrete masses of organic material are deposited, the so-called mammillae, whose surface contain mammillan, a calcium-binding keratin sulfate-rich glycoprotein, which is involved in the nucleation of the first eggshell calcite crystals [172–174]. Another glycoprotein, which accompanies the growth of the crystalline columns of calcite, is ovoglycan, a dermatan sulfate-rich glycoprotein [172–175].

Summarizing, most biomineralization proteins have a multidomain structure and can often be glycosylated by hydrophilic and often acidic polysaccharides, which also can form the domains interacting with the mineral.

3.2
Bio-inspired Mineralization

The term bio-inspired mineralization can be understood in a broad sense. For example, the above mentioned retrosynthesis of nacre [132] is as much bio-inspired as is the crystallization in presence of hydrophilic polymers [44], although in the first case, the original biomineral structure is reproduced, whereas in the latter case, no biomineral is reproduced but complex crystal morphologies are obtained. The inspiration of both examples by biomineralization is the application of polymer additives to modify the crystallization of a mineral. This can in turn even enable a closer understanding of natural biomineralization processes, as in the case of the amorphous precursors for nacre retrosynthesis. As discussed above, the structural motifs of biomineralization polymers are manifold, so that a general design strategy for a polymer, which mimics the biomineralization polymers is not possible. Structures, which more resemble polyanions (polysaccharides) are also found as structures with blocked and hydrophilic blocks (dentin phosphoryns and sialoproteins). Consequently, suitable bio-inspired crystallization additives can be polyelectrolytes and block or graft copolymers with hydrophilic and acidic blocks (DHBCs and DHGCs). But also, more complex functionalization patterns can be applied in bio-inspired mineralization, like for example designed peptides from combinatorial approaches as well as the biomineralization polymers directly. Thus, in the broadest sense, bio-inspired mineralization can be understood as the application of soluble polymeric additives (mimics of the soluble matrix) for the control of crystallization reactions.

Similar considerations apply for the insoluble matrix, which can be mimicked by various structured templates acting as a mould for the forming mineral. Only in exceptional cases (like in the nacre retrosynthesis example [132]) was it possible to obtain an organic/inorganic composite material, which was indistinguishable in structure from the original biomineral by electron microscopy. However, the principal limitation of the current bio-inspired bottom-up mineralization approaches is that they can only yield self-assembled structures up to the micrometer level, unless an external template is provided like in the nacre retrosynthesis case. Higher structural levels that can be found in biominerals are controlled by cell action, which so far does not have a close synthetic mimic. Therefore, only structured templates like patterned monolayers [176, 177], biomineral replicas [178, 179], or the original structural biomineral matrix [132] can be applied to achieve a structuration up to the macroscopic scale. Nevertheless, despite these limitations, the study of bio-inspired mineralization processes and the investigation of the complex crystallization and self-assembly processes make sense. They can help to understand parts of the biomineralization processes and explore ways in which biomineralization principles can be transferred to the non-living world to allow for the synthesis of advanced future materials.

3.2.1
Biopolymers

The application of biopolymers as mineralization additives is a natural bioinspired choice due to their role in biomineralization processes. However, biomineralization processes are very complex and usually involve a structural (insoluble) and functional (soluble) matrix. Commonly, the structural matrix is neglected and investigations focus on the soluble crystallization additives.

In vitro experiments with biopolymer additives have so far commonly failed to reproduce the complex biomineral shapes. Examples include $CaCO_3$ mineralized in the presence of cationic, anionic and non-ionic dextran [138], collagen [180], plant or bacterial enzymes [181], soluble mollusk shell proteins extracted from nacre [182], as well as the stabilization of spherical ACC by extracted soluble macromolecules from coralline algae [183], protein secondary structures for the control of the spatial orientation of functional groups so that the protein can bind to a targeted crystal face [92], as well as a combinatorial approach to select peptides from a 15-mer peptide library as additives for $CaCO_3$ crystallization [184]. The failure of the above examples to reproduce complex biomineral shapes can be mainly attributed to the complicated biomineralization process, where an insoluble matrix also influences the crystallization location as a compartment, so that the pure soluble matrix – or even only a part thereof – cannot alone remodel the complex crystal shape. Also, possible synergetic interactions between the various components in the organic matrix of a forming biomineral are difficult to mimic. Nevertheless, the morphologies of the crystals obtained in the above examples are still modified, often to a large extent. In addition, ordered self-assembly can be induced, as observed for hydroxyapatite mineralized in the presence of the glycosaminoglycan chondroitin-4-sulfate [185].

Not only the morphology but also the polymorph can be influenced by soluble biopolymers. A prominent example of the influence of soluble biopolymers on a crystallization event is the control of $CaCO_3$ crystal modification (aragonite/calcite) in the presence of soluble macromolecules extracted from the respective layers of a mollusk shell [50, 85], as well as the $CaCO_3$ morphology [182] (for a review, see [148]). Also, proteins can promote the crystallization of a preferred mineral as was shown for the promotion of fluorapatite crystallization by soluble matrix proteins from *Lingula Anatina* shells [186].

Biopolymers can offer specific interactions, which can be exploited to achieve a self-assembly or molecular recognition and by this means structure inorganic matter. In that respect, the use of DNA in conjunction with inorganic materials in nanostructure assembly has been well demonstrated due to the strong electrostatic and coordinating interaction of DNA with nanoparticles including Au [187–191], Ag [192, 193], Pd [194], Pt [195] CdS [196, 197], and CdSe [198]. The programmed self-assembly of Au nanoparticle assemblies has been confirmed by the interaction of DNA with Au nanoparticles or nanorods to form either small two-dimensional structures, or large three-dimensional structures [187–200], including liquid crystals formed via alignment of virus-bound Au nanoparticles [201]. The duplex linkages between complimentary surface-anchored oligonucleotides especially provide a nice possibility for programmed particle self-assembly [200]. A similar approach using DNA double-stranded crosslinks with oligonucleotides bound to the surfaces of Au and ferritin nanoparticles allowed for the self-assembly of disordered networks between Au and ferritin, which could be reversibly disassembled by heating [202]. The same strategy could be used to bind Au nanoparticles onto the surface of mesoporous silica particles [203].

Besides DNA, other linear fibrous biopolymers like fibrin, dextran, and collagen can be used for the coordination of nanoparticles to produce linear nanoparticle arrays. Examples include Au and Pd nanoparticles assembled on peptide nanofibers [204, 205], various metals and metal oxide nanoparticles on a dextran template [206], TiO_2 and SnO_2 nanoparticles forming a one-dimensional array on a collagen template [207] as well as ZnS, CdS, CoPt, and FePt nanoparticle chains assembled on bacteriophage [201, 208–210]. The approach of nanoparticle coordination by biomolecules was also extended to other proteins with the aim of protein wrapping to produce functionally isolated hybrid nanoparticles like myoglobin, haemoglobin, or glucose oxidase wrapped with an ultrathin shell of cationic aminopropyl-functionalized clay oligomers [211]. Exfoliated sheets of the clay compound also interacted with negatively charged biomolecules resulting in mesolamellar organic/inorganic hybrid structures under preservation of the protein secondary structures for myoglobin and glucose oxidase [212]. The wrapped enzyme showed activity even at higher temperatures and over an extended pH range. Proteins could also be coated with an amorphous silica shell without loss of secondary structure and showing an increased thermal stability [213].

DNA was also used as a template molecule for $BaSO_4$ crystallization. Unusual needle shaped microcrystals with secondary perpendicular grown needles were observed, at high $BaSO_4$ supersaturation, whereas at a moderate supersaturation, flower-like crystals similar to that shown later in Fig. 22 were obtained, clearly demonstrating the DNA influence [214]. Other linear biopolymers like dextran or dextran sulfate could also be successfully applied as a crystallization additive rather than a template for nanoparticle coordination, as demonstrated for silver synthesis in the presence of these molecules resulting in spherical and various plate morphologies [215]. On the other hand, a viscous dextran solution can also serve as sacrificial template for preformed nanoparticle arrangement with subsequent thermal decomposition of the dextran, as demonstrated for zeolite/silica or NaY/silica frameworks [216], or metal and metal oxide sponges [217], where the nanoparticles aggregate and coalesce into interconnecting filaments between gas bubbles formed by dextran decomposition. Fibers drawn from a viscous dextran solution with silicalite nanoparticles led to aligned silicalite nanocrystal fibers, with subsequent removal of the dextran template [216]. A similar approach, but with a dextran sulfate solution template and amino-acid coated hydroxyapatite nanoparticles, could be used to generate hydroxyapatite sponges, which showed promising properties for cartilage and soft tissue engineering [218]. Elevated temperatures were also applied in a hydrothermal approach using starch and other polysaccharides for the production of carbon-coated silver nanocables with remarkably uniform core and shell diameters and interconnection points [219]. Silver ions are reduced by starch leading to silver particles entrapped in a starch gel matrix with subsequent aggregation of several gel strings. Then, the silver-catalyzed carbonization reaction of the starch takes place, finally leading to the observed silver nanocables shown in Fig. 11.

Biopolymer self-assembly can be achieved by complexation of multivalent ions. One such example is the self-organized microgel reactor, which is formed from Fe^{3+} and κ-carrageenan double helices in aqueous solution [220]. This self-assembled structure could be used for the synthesis of stable $FeO(OH)_2$ nanoparticles in the interior [220] as well as for other metal hydroxides [221].

As one kind of amazing biosilica-producing organism, diatoms exhibit complicated and highly organized superstructures that exist in living organisms. Recently, Kröger et al. reported that the complex morphology of the nanopatterned silica diatom cell walls is related to species-specific sets of polycationic peptides, so-called silaffins, which have been isolated from diatom cell walls [154]. Silica nanospheres and their networks can form in vitro within seconds when silaffins are added to a solution of silicic acid, implying the catalytic role of the silaffins for the silica polycondensation. The morphology of precipitated silica particles can be controlled by changing the chain length of the polyamines as well as by a synergetic action of long-chain polyamines and silaffins [222]. The results imply that similar mixtures are

Fig. 11 **a** SEM image of the nanocables with encapsulated pentagonal silver nanowires. The *inset* shows an incompletely filled tube with pentagonal cross section. **b,c** TEM micrographs showing various forms of the nanocables. Figure reproduced from [219] with permission of Wiley

used in vivo by the diatoms to create their intricate mineralized cell walls. Sumper has recently developed a model to explain the delicate pattern formation in diatom shells by phase separation of silica solutions in the presence of these polyamines [223]. This demonstrates that the role of a biomineralization polymer can be manifold – in this case, it acts as a catalyst and also as a mould for silica condensation by induced phase separation.

More recently, the same group has developed a gentle extraction method for the extraction of silaffins in their native state in order to avoid the use of harsh anhydrous HF treatment for dissolving biosilica during extraction. The extracted silaffins contain additional modifications besides their polyamine moieties [224]. The phosphorylation of the serine residue is essential for biological activity and a plastic silaffin–silica phase was detected through time-resolved analysis of silica morphogenesis in vitro, which could represent building materials for diatom biosilica [224]. A 19 amino acid peptide derived from silaffin showed the ability to form several silica morphologies ranging from spheres to complex fibrils [225].

Various linear synthetic analogs of the natural active polyamines in biosilica formation on the basis of linear poly(ethylene imine) or poly(propylene imine) were recently synthesized and reported to accelerate the silic acid condensation even more than the above-mentioned silaffins [226]. This implies that it is not the silaffin protein superstructure that is responsible for its catalytic activity, as is the case for example in enzymes. In addition, poly(L-lysine) also led to modified silica morphologies such as fibers or ladders [227].

3.2.2
Homopolymers and Random Copolymers

Homopolymers and random copolymers are a naturally simple choice from the toolbox of synthetic polymers, as many of these polymers are commercially available and do not need to be synthesized by the researcher. The electrosteric stabilization of inorganic colloids by polyelectrolytes has been demonstrated for iron oxyhydroxide, which was immediately stabilized by cellulose-sulfate and κ-carrageenan to avoid the precipitation of amorphous solid [220]. In addition to stabilization, the use of low molecular mass polyelectrolytes such as polyacrylic acid or polyaspartic acid crystal growth modifiers or structure-directing agents can lead to structured calcium phosphate [228] or the formation of unusual hierarchical superstructures such as helical $CaCO_3$ [137, 229], complex spherical $BaCO_3$ superstructures [230], hollow octacalcium phosphate ($Ca_8H_2(PO_4)_6 \cdot H_2O$) [231], $BaSO_4$ [109] and $BaCrO_4$ fiber bundles or superstructures with complex repetitive patterns [111]. It has to be pointed out that the experimental window for the formation of the latter superstructures is narrow.

Unusual complex structures of calcite helices or hollow helices can be obtained via PILP liquid precusors in the presence of a chiral as well as achiral poly(aspartate). This can produce helical protrusions or occasionally produce hollow helices as observed by Gower et al. [137], showing that it is not essential that the polypeptide has to be chiral to produce these helical structures (Fig. 9). Very recently, a $CaCO_3$ PILP with PAsp was used to infiltrate and mineralize a PHEMA hydrogel replica of a sea urchin spine [232]. Although the structural replication of the spine was not perfect, this example shows that PILPs can be even used to enter pores within a gel structure.

The crystallization via PILP precursors proved to be very versatile for the generation of complex crystalline morphologies. So far, the PILP process has been mainly established for $CaCO_3$ systems. Interestingly, PILP processes could also be revealed for organic crystals, as recently demonstrated for various acidic and basic amino acid systems that formed PILPs upon addition of the counstercharge polycation or polyanion (PAA and PEI) [141]. For this system, a direct proof of the liquid PILP character was possible by centrifugation and observation of a very viscous PILP phase (Fig. 12c). Crystallization of the PILP droplets resulted in porous microspheres composed of nanoplates

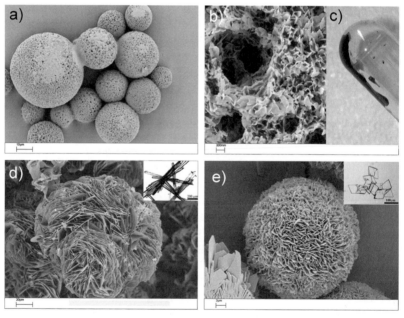

Fig. 12 **a** DL-Glutamic acid crystals obtained from a concentrated PILP precursor phase with PEI. **b** Zoom in of **a** showing the porous structure and the nanoplate building units. **c** Experimental proof of the liquid character of a DL-glutamic acid PILP phase after centrifugation and removal of the supernatant aqueous phase. **d** L-Histidine crystals obtained from a concentrated PILP precursor phase with PAA. **e** L-Lysine crystals obtained from a concentrated PILP precursor phase with PAA. The *insets* show the default crystals without additives. *Scale bars* represent **a** 10 μm, **b** 200 nm, **d** 20 μm, and **e** 3 μm. Figure reproduced from [141] with permission of Wiley

(Fig. 12). These amino acid spheres have a size suitable for packing in chromatographic columns and are available in an enantiomeric form so that they may find potential application in chromatographic racemate separations as they are much easier to obtain than the imprinted particles applied up to now.

However, the most intensively studied system with liquid precursors is still $CaCO_3$. If PILP droplets are deposited on a substrate, they coalesce and form a coating, which subsequently transforms into patchwork-like calcite films with different single crystalline domains via an amorphous to crystalline transition [229].

Gower et al. were also able to synthesize calcite nanofibers via a PILP process, in a deposition mechanism termed solution–precursor–solid (SPS) mechanism, which has features of vapor–liquid–solid and solution–liquid–solid deposition mechanisms [233]. In these experiments, calcite seed rhombohedra were overgrown with fibers, with rounded tips indicating their growth via a liquid precursor. Compared to the high temperature vapor deposition techniques, the PILP route is operational at ambient temperatures down to 4 °C, which indicates a promising route towards fibrous materials. $CaCO_3$ PILPs were also found to mineralize collagen and amazingly, the PILP precursors were even able to enter the nonometer-sized gap zones of collagen resulting in a structure partly resembling that of bone [234]. These results suggest that bone mineralization might also proceed via a PILP precursor and a paradigm was suggested that Ca-based biominerals form via PILP precursors [145] although the current evidence is still based on $CaCO_3$ only.

If the PILP technique is used to deposit an amorphous precursor film, which is composed of ACC nanoparticles and PAA on a substrate, it can be slowly transformed into a partially crystalline film by slow drying [235]. Heating yields a crystalline film with a patchwork structure of single crystalline patches with differing crystallographic orientations, as revealed by polarization microscopy (see also Fig. 13). Overgrowth of the patchwork film with $CaCO_3$ without polymer resulted in single and multiple layers of highly oriented calcite crystals grown on top of the polycrystalline patchwork film, finally leading to a laminated $CaCO_3$ coating (Fig. 13). The crystal texture changes abruptly at the borderline between different domains but stays constant within each individual domain, supporting the epitaxial overgrowth. The multilayered structure is reminiscent of the nacre architecture although the present films have a simpler architecture. Nevertheless, they show iridescence so that the presented route is an interesting low cost and bio-inspired pathway towards photonic materials [235].

A close structural mimic to nacre, with aragonite tablets being a mesocrystal composed of aligned crystalline nanoparticles [115], was reported to form spontaneously from a precursor solution containing K_2SO_4 and PAA [115]. As found for the nacre archetype, the K_2SO_4–PAA composite is composed of platy units with the architecture of the layered units being an oriented assembly of nanoparticle units oriented along the b-axis and scattering like

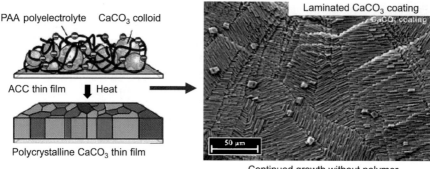

Fig. 13 Three-step procedure for the morphosynthesis of nacre-type laminated CaCO₃ coatings. In the first step, an amorphous highly hydrated CaCO₃ thin film is deposited on a glass substrate. Upon heating, this precursor film is transformed into a polycrystalline thin film consisting of a mosaic of flat single-crystalline calcite domains. In the last step, highly oriented single and multiple layers of calcite crystals are grown epitaxially on the underlying polycrystalline thin film. Reproduced from [235] with permission of Wiley

a single crystal, in contrast to the c-axis orientation for natural nacre. Also, the K_2SO_4–PAA composite showed nanostorage properties like natural nacre, allowing for the infiltration of dye molecules [115]. Thus, the reported purely synthetic system is already a close structural mimic to natural nacre exhibiting all the important features on several hierarchy levels (see also Fig. 15).

Complex spherical shell architectures of $Ca_8H_2(PO_4)_6 \cdot 5H_2O$ (OCP) precipitated from aqueous solutions containing polyaspartate or polyacrylate were explained by a similar mechanism as that proposed for the CaCO₃ helices in Fig. 9 by Gower et al. [231], namely that the hollow microstructures consist of a thin porous membrane of oriented OCP crystals that are highly interconnected. The hierarchical growth processes involve the radial growth of dense, multilayered spheroids, secondary overgrowth of a porous thin-shell precursor, and anisotropic dissolution of the spheroidal cores, producing this remarkable hollow structure [231]. A double jet precipitation of calcium phosphates in the presence of the related polymer sodium polyaspartate on the other hand results in the formation of hierarchical structures of calcium phosphates through structure transformation from amorphous precursors into well-defined structures [228].

In addition to the above reported helical or hollow spherical morphologies, elegant nanofiber bundles and their superstructures of inorganic crystals could be generated by the process of oriented attachment (Fig. 14). The sodium salt of polyacrylic acid can serve as a very simple structure-directing agent for the large scale, room temperature synthesis of highly ordered cone-like crystals [109] or very long, extended nanofibers of $BaCrO_4$ or $BaSO_4$ with hierarchical and repetitive growth patterns [111]. Temperature and concentration variation allow control of the finer details of the architecture, namely

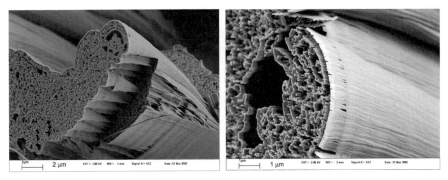

Fig. 14 Complex forms of $BaSO_4$ fiber bundles produced in the presence of 0.11 mM sodium polyacrylate (M_n = 5100), at room temperature, [$BaSO_4$] = 2 mM, pH 5.5, 4 days. Reproduced from [151] with permission of the Materials Research Society (Fig. 4c,d)

length, axial ratio, opening angle, and mutual packing [109]. The observed (210) growth axis implies that the polyanion adsorbs to all parallel faces to this axis on the nucleated nanoparticles leaving the negatively charged (210) faces uncovered and thus making them the highest energy faces, which then fuse together by oriented attachment to form the fibers (see also Fig. 25 and corresponding discussion).

The formation of interesting hierarchical and repetitive superstructures by simple polyelectrolyte additives is worth exploring further in other mineral systems. Other low molecular weight polyelectrolytes, such as poly(allylamine hydrochloride) (PAH) and poly(sodium 4-styrenesulfonate) (PSS), can be used for the self-assembly of complex spherical $BaCO_3$ superstructures through a facile mineralization process under ambient conditions [230]. The obtained $BaCO_3$ spherulites were built from smaller, elongated rod-like building blocks with a typical diameter of 50 nm and length of 200 nm, which apparently adopted the more equilibrated isostructural aragonite appearance. Crystallization of spherical $CaCO_3$ particles in the presence of PSS has also been reported by Jada et al. [236] and spherical, rod-like and flower-like $BaSO_4$ particles [237] as well as hierarchically structured porous $CaCO_3$ microspheres [238] and monodisperse peanut shaped celestine ($SrSO_4$) particles with mesoporous microstructure [239] were obtained with a PSMA additive. On the other hand, PAM was found to be a successful additive for promoting the formation of well-crystallized aragonite ($CaCO_3$) and $BaCO_3$ rods at elevated temperatures [240].

The addition of polyelectrolytes with strong inhibition ability [241, 242] will stabilize amorphous nanobuilding blocks in the early stage and then stimulate a mesoscale transformation [40] or act as a material depot in a dissolution–recrystallization process. This could be shown in a time resolved study on the $CaCO_3$ scale inhibition efficiency of polycarboxylates, where amorphous precursor particles were detected in the initial stages [243].

The additive systems for crystallization control can also get more complex than the discussed single polymer additives. For example, Deng et al. have recently reported a ternary additive system of PEG, PMAA, and SDS for the crystallization control of $CaCO_3$ [244]. Here, the active additive is formed by a PEG–PMAA complex and the simultaneous PEG–SDS interaction to form complex micelles [245]. A variety of nanoparticle aggregate structures like spheres or hollow spheres could be obtained. This approach is interesting because it uses simple and commonly available components instead of the specialized block copolymers, which essentially yield similar results for these particular polymer blocks. Also, the additive system is modular, allowing for a large variety of additive variations exploiting the interaction and cooperation between soluble macromolecules, which can create synergetic and cooperative effects [244].

Another recent work by Tremel et al. addressed the common limitation of the bio-inspired mineralization approaches in only using a mimic of the soluble matrix in biomineralization even though the structural matrix also plays a role. Tremel applied a self-assembled monolayer as mimic for the structural matrix and PAA as the soluble matrix, which adsorbs onto the SAM [246]. ACC nanoparticles attach to the PAA/SAM and subsequently, vaterite nanowires are formed by cooperative PAA/SAM action. These nanowires are themselves aggregates of nanoparticles that nucleated from the ACC nanoparticles on the PAA backbone, which initially formed a wire template by crosslinking of several PAA chains by Ca^{2+}. The strategy of a combined application of a soluble and insoluble matrix was also applied in the nacre retrosynthesis approach using the demineralized organic nacre matrix as the insoluble and polyaspartic acid as the soluble matrix ([132], and Fig. 8). As also found by Tremel [246], application of functional and structural matrices together led to different results than could be obtained by the application of a single additive. In the nacre retrosynthesis example, a material indistinguishable from the natural archetype by electron microscopy could be obtained [132] (Fig. 8). The nucleation of an amorphous $CaCO_3$ film under a Langmuir monolayer (mimic of insoluble matrix) in the presence of PAA as mimic of the soluble matrix, observed by in situ X-ray scattering techniques, was also reported and showed that the nucleation of the film is not directed by templating on an atomic scale [247].

These results clearly show that most of the current bio-inspired mineralization approaches use over-simplified systems, which is a result of the often cumbersome analytics of the obtained organic/inorganic hybrid systems and their formation mechanisms.

Recently, chiral copolymers of phosphorylated serine and aspartic acid with molar masses between 15 000–20 000 g mol^{-1} were reported to be very efficient additives for the generation of helical calcite superstructures consisting of elongated 70 nm wide, uniform and highly aligned calcite nanoparticles where the helix turn corresponded to the copolymer enantiomer [248] (Fig. 15).

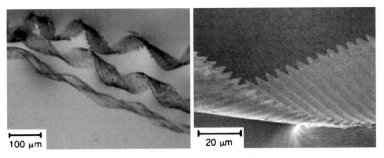

Fig. 15 Optical and SEM micrographs of chiral morphologies of $CaCO_3$. Helical growth in the presence of a copoly[L-Ser-(P)$_{75}$L-Asp$_{25}$], a clockwise P twisted spiral morphology. Reproduced from [248] with permission of Wiley

For the growth of the helical structures, a narrow experimental window was identified. The helical structures formed when a high degree of phosphorylated Ser (75 mol%) and 25 mol% Asp in the copolymer were applied in combination with a tenfold Ca^{2+} concentration with respect to the monomer – similar to the conditions where a shellfish forms a shell. Nevertheless, the formation mechanism of the chiral crystalline superstructure is unclear and remains to be revealed in the future.

Helices from twisted twin subunits were reported for triclinic $K_2Cr_2O_7$ and H_3BO_3, formed in a diffusion field created around a growing crystal in poly PAA or PVA gels [249] or orthorhombic K_2SO_4 in a viscous PAA solution [250]. In the latter case, the crystal habit was modified by selective PAA adsorption leading to tilted platy crystals in addition to the diffusion-limited growth condition, controlling the assembly of the tilted subunits. Even the direction of the helical turn could be influenced for $K_2Cr_2O_7$ in a gel medium by molecular recognition of the enantiomeric triclinic subunit surfaces by glutamic acid enantiomers [251].

The face-selective PAA adsorption onto orthorhombic K_2SO_4 crystals led to the formation of tilted unit crystals, which were assembled in a diffusion-limited condition resulting in various complex morphologies like helices or zig-zag assembly of twinned crystals [250]. A conclusive explanation of the various possibilities of particle growth in an anisotropic diffusion field was given. It is remarkable that the K_2SO_4–PAA system has six hierarchical levels from the nanometer scale to that of several hundreds of micrometers, which is a typical feature of biominerals and has so far only ben rarely reported for a synthetic system (Fig. 16) [252].

Superstructure design at each level was controllable by changing the polymer concentration and the observed hierarchy was attributed to the interaction between crystals and polymers and the diffusion-controlled conditions [252]. A similar hierarchical system was recently found for potassium hydrogen phthalate and PAA [253]. Again, plate-like units were composed of

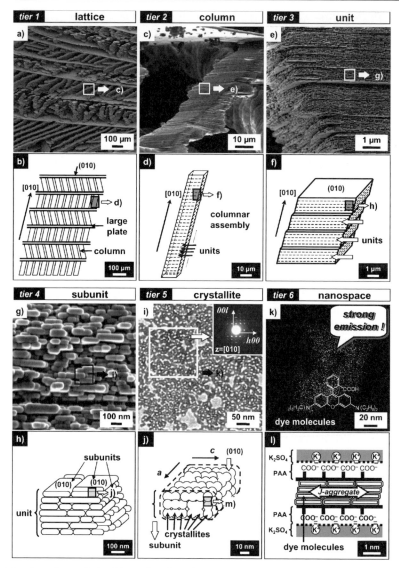

Fig. 16 Overview of the hierarchical architecture in the K_2SO_4–PAA system and its schematic illustrations at six different size scales ($c_{PAA} = 10$ g L^{-1}): **a,b** field emission scanning electron microscopy (FESEM) image and a schematic representation of the macroscopic lattice architecture consisting of large thin plates and columns (*Tier 1*); **c,d** columnar assembly between plates (*Tier 2*); **e,f** units in the columns (*Tier 3*); **g,h** subunits inside a unit (*Tier 4*); **i,j** field-emission transmission electron microscopy (FETEM) image and schematic representation of crystallites with the same orientation in a subunit; *inset* corresponding selected area electron diffraction (SAED) patterns taken along the [010] direction (*Tier 5*); **k,l** energy-filtered TEM (EF-TEM) image and schematic illustration of the organization of dye molecules in a nanostorage space (*Tier 6*). Figure reproduced from [252] with kind permission of Wiley

aligned crystalline nanocrystals, so that the well-facetted plates on the micrometer scale can be considered as mesocrystals, although the authors of the original paper also discuss mineral bridges between the subunits as a possibility for the explanation of their mutual crystallographic alignment [253]. As the concept of particle assembly in diffusion fields coupled with face-selective polymer adsorption was demonstrated for both inorganic and organic crystals [252, 253], it seems to be much more versatile than realized so far.

Other mesocrystal systems were also recently reported. $CaCO_3$ precipitation in the presence of a PSS additive yielded mesocrystals with a morphology that could be systematically changed by variation of the $PSS/CaCO_3$ ratio [122]. At higher $CaCO_3$ concentrations, even the symmetry of the morphology was broken and convex–concave shaped mesocrystals with a central hole were obtained (Fig. 17).

The observed morphogenesis sequence in Fig. 17 was speculatively explained by the generation of nanoparticles with oppositely charged counter

Fig. 17 Typical SEM images of calcite mesocrystals obtained on a glass slip by the gas diffusion reaction after 1 day in 1 mL of solution with different concentrations of Ca^{2+} and polystyrene-sulfonate: **a** $[Ca^{2+}] = 1.25\,\text{mmol L}^{-1}$, $[PSS] = 1.0\,\text{g L}^{-1}$; **b** $[Ca^{2+}] = 1.25\,\text{mmol L}^{-1}$, $[PSS] = 0.5\,\text{g L}^{-1}$; **c** $[Ca^{2+}] = 1.25\,\text{mmol L}^{-1}$, $[PSS] = 0.1\,\text{g L}^{-1}$; **d** $[Ca^{2+}] = 2.5\,\text{mmol L}^{-1}$, $[PSS] = 1.0\,\text{g L}^{-1}$; **e** $[Ca^{2+}] = 2.5\,\text{mmol L}^{-1}$, $[PSS] = 0.5\,\text{g L}^{-1}$; **f** $[Ca^{2+}] = 2.5\,\text{mmol L}^{-1}$, $[PSS] = 0.1\,\text{g L}^{-1}$; **g** $[Ca^{2+}] = 5\,\text{mmol L}^{-1}$, $[PSS] = 1.0\,\text{g L}^{-1}$; **h** $[Ca^{2+}] = 5\,\text{mmol L}^{-1}$, $[PSS] = 0.5\,\text{g L}^{-1}$; **i** $[Ca^{2+}] = 5\,\text{mmol L}^{-1}$, $[PSS] = 0.1\,\text{g L}^{-1}$. Figure partly reproduced from [122] with permission of the American Chemical Society

(001) faces as a result of dielectric interaction throughout the crystal after adsorption of the first PSS molecules on one (001) face. This leads to dipole fields, which can order and mutually align the nanocrystalline building units of the mesocrystal; a similar proposal to the one initially made for fluorapatite growth in a gelatin matrix [117].

Face-selective polymer adsorption can also lead to the formation of one-dimensional particle aggregate arrays without the particle fusion of the individual nanoparticles that takes place during oriented attachment. This was demonstrated for silicalite nanoparticles, which self-assembled to chains after selective PDADMAC adsorption [254].

Even very simple polymers without an obvious functionality can have an influence on a mineralization process, as demonstrated for the generation of porous silica in the presence of PEG [255]. PEG had a multiple role serving as flocculation agent in silica sols, silica polymerization agent, phase separation agent, and porogen to generate pore dimensions of 2–20 nm.

The knowledge gained through the hydrothermal synthesis of silver nanocables in the presence of starch [219] could be extended to synthetic homopolymers and other inorganic systems. Examples are silver nanocables with crosslinked PVA shells, where the PVA has synergetic effects as stabilizer and for silver binding [256], or copper ions resulting in PVA-coated copper nanocables with lengths up to $100\,\mu m$ [257]. The latter study demonstrated that even multivalent ions are accessible by this easy large scale synthesis approach leading to uniform and flexible nanocables.

3.2.3
Dendrimers

Dendrimers have been widely used as organic matrices for the synthesis of inorganic materials as a mimic of micelles or proteins [258–270]. Highly soluble nanocomposites of ferrimagnetic iron oxides and dendrimers are produced with good stability under a wide range of temperature and pH [261]. Recently, dendrimers were discovered as active additives for controlled $CaCO_3$ precipitation [271], as reviewed by Naka [272]. A motivation for this is the defined dendrimer structure, shape, and chemical functionality. Anionic starburst dendrimers were found to stabilize spherical vaterite particles for up to a week with controllable particle size in the range of 2.3–5.5 nm, depending on the dendrimer generation number [273]. Poly(propylene imine) dendrimers interacting with octadecylamine are the only known synthetic polymeric species that are reported to stabilize kinetically formed amorphous calcium carbonate (ACC) for periods exceeding two weeks in water [274, 275]. The latter system appears especially versatile, as the dendrimer functionality can be easily varied by a simple exchange of the surfactant hydrophobically interacting with the dendrimer.

3.2.4
Double Hydrophilic Block Copolymers (DHBCs)

Following the classical understanding of the term "amphiphilic", block copolymers usually consist of a hydrophilic and hydrophobic block so that they can be used as polymeric surfactants, which show a similar behavior to low molecular weight surfactants [276, 277]. However, the term amphiphilic (Greek for "loving both") can be understood in the much more general sense of an affinity to two different phases, regardless of what they are. Most important for bio-inspired mineralization is certainly a mineral phase in an aqueous environment. To address these phases in a targeted way, a new class of functional polymers, the so-called double-hydrophilic block copolymers (DHBCs), has been designed for mineralization purposes [147]. These polymers are bio-inspired insofar as they just contain those parts of a protein that are commonly recognized as those interacting with a mineral surface (acidic or basic amino acid units), and those which keep the protein dissolved in the aqueous environment where biomineralization takes place (hydrophilic amino acid residues). As active biomineralization proteins often contain high amounts of acidic amino acids in blocked sequences, the DHBC design is a blocked structure serving as a model system where the block lengths can be varied to study the influence on crystallization events.

Consequently, DHBCs consist of one hydrophilic block designed to interact strongly with the appropriate inorganic minerals and surfaces, and another hydrophilic block that does not interact (or only weakly interacts), and mainly promotes solubilization in water (Fig. 18). Owing to the separation of the binding and the solvating moieties, DHBCs are "improved versions" of polyelectrolyte homopolymers (see Sect. 3.2.2) and turn out to be extraordinarily effective in crystallization control of various minerals. These polymers are typically rather small, having block lengths of 10^3–10^4 g mol^{-1}. This has also practical reasons, as too long an interacting block might stick to more than one mineral nanoparticle and would thus lead to particle aggregation and bridging flocculation [278]. In addition, longer interacting blocks may lead to aggregation of the DHBC upon interaction with bivalent ions like Ca^{2+} [279].

Using the DHBC design, not only the required lowering of the crystal–water interface energy by polymer adsorption is obtained, but it is also possible to create a high specificity to distinct crystalline surfaces by adjustment of the functional polymer pattern [280].

Due to the DHBC design, the stabilizing block can be tailored in length so that either steric particle stabilization (a in Fig. 18) or only temporary, steric nanoparticle stabilization is achieved. This leads to nanoparticle building units for complex morphologies. The short stabilizing block lengths of typical DHBCs of 2000–5000 g mol^{-1} are in the range for at least temporary steric particle stabilization but short enough to still enable particle

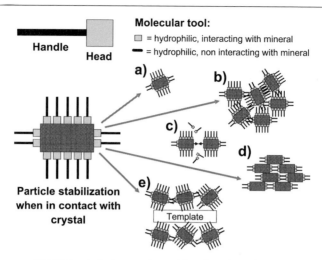

Fig. 18 Concept of DHBCs, their interaction with mineral surfaces and possibilities for particle stabilization or self-assembly. *a* Nanoparticle stabilization, *b* unordered particle aggregation, *c* oriented attachment, *d* ordered aggregation/mesocrystal formation, *e* arrangement of nanoparticles around a template. Reproduced from [151] with permission of the Materials Research Society

self-assembly [278]. This can be unordered nanoparticle aggregation (b in Fig. 18), oriented attachment of nanoparticles by DHBC displacement or face-selective DHBC adsorption (c in Fig. 18), ordered programmed self-assembly by face-selective DHBC adsorption (d in Fig. 18), and arrangement of temporary stabilized nanoparticles around a template (e in Fig. 18).

From the viewpoint of particle stabilization, DHBCs have an optimized molecular design combining the advantages of electrostatic particle stabilization with those of steric particle stabilization by polymers. In addition, they are able to selectively adsorb on certain crystal faces and can therefore control particle shape upon further growth. This can also be achieved with ions or low molar mass additives (Fig. 19a) but then the particle is not sterically stabilized. On the other hand, polymers can be adsorbed and stabilize the particle (Fig. 19b), but if the polymers are long, they do not adsorb selectively but cover all crystal faces. A DHBC, however with a short sticking block combines the advantages of face-selective adsorption with that of particle stabilization due to the longer stabilizing block (Fig. 19c).

The proper adjustment of the block lengths between stabilizing and interacting block is also an important issue in DHBC design. It appears plausible that a DHBC with too short an interacting block will not stick well on a particle surface. On the other hand, too long a stabilizing block has too high space demands, so that less DHBCs can adsorb on a particle surface and the stabilization effect is decreased. This is outlined in Fig. 20. In Fig. 20a, the sticking block is too long and the stabilizing block too short so that the stick-

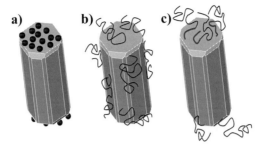

Fig. 19 Face-selective adsorption of ions or low molar mass additives (**a**), steric particle stabilization by polymers (**b**), and face-selective adsorption and particle stabilization by DHBC (**c**). Figure reproduced from [57] with permission of the Royal Society of Chemistry

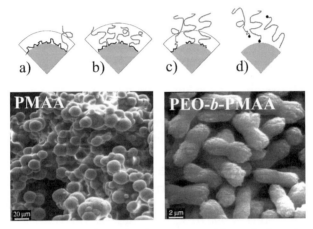

Fig. 20 Stabilization of a particle surface by a DHBC with a polyelectrolyte (PE) sticker block and a stabilizing (ST) block to illustrate the necessity of an appropriate block length ratio. **a** PE block too long and ST block too short, **b** PE and ST block in good ratio, **c** PE block too short, ST block too long, and **d** PE block far too short and ST block too long. The SEM micrographs illustrate the difference of a PMAA polymer additive as homopolymer and DHBC with stabilizing PEO block for the example of $CaCO_3$ crystallization. Reprinted from [282] with permission of Wiley

ing block demands too much space on the particle surface, leading to a low polymer coverage and bad stabilization. In extreme cases of polymer adsorption onto several particle surfaces it can even lead to bridging flocculation. If the length ratio between stabilizing and sticking block is balanced, a good particle stabilization can be achieved (Fig. 20b) whereas too long a stabilizer block also leads to a bad particle stabilization because its space requirements lead to a low polymer coverage of the particle (Fig. 20c). In an extreme case, the sticker block is far too short and the DHBC does not adsorb on the particle surface anymore. These considerations were recently verified experimentally for the stabilization of ACC by PEO-*b*-PAA [281].

In typical DHBCs for mineral stabilization, the polyelectrolyte sticking block is 500–5000 g mol^{-1} whereas the stabilizing block is 2000–10 000 g mol^{-1}. According to the above considerations, the length of the stabilizer block can be used to tune particle stabilization, so that mineral nanoparticles with only a temporary stabilization can be also generated. These can be used as building units for complex superstructures via a self-assembly mechanism. The advantage of block copolymers as compared to homopolymers for the stabilization of particles is shown for the example of CaCO$_3$, which was precipitated in the presence of a PMAA homopolymer and a PMAA DHBC with a stabilizing PEO block (Fig. 20). For the homopolymer, spherical particle aggregates are obtained, which are glued together by PMAA molecules adsorbing onto several nanoparticle surfaces. This does not happen for the DHBC, as the stabilizing PEO shell prevents gluing of the dumbbell-shaped microparticles, which remain dispersed.

DHBCs have shown remarkable effects in synthesis, stabilization, and crystallization for a wide range of inorganic, inorganic/organic hybrids, as well as organic crystals with interesting shapes and complex forms through the mesoscale self-assembly of the "programmed" primary units (for a recent review see [57]). They can therefore nicely serve as a bio-inspired soluble matrix for controlled particle self-assembly.

Different kinds of DHBCs with different functional patterns have been designed and used as crystal modifiers [57, 147]. All functional groups, which are important for the interaction with inorganic crystals, were synthetically realized. More specificly, DHBCs reported to date have functional blocks with: – OH, – COOH, – SO$_3$H, – SO$_4$, – PO$_3$H$_2$, – PO$_4$H$_2$, – SCN, – NR$_3$, – HNR$_2$, and – H$_2$NR (for a review on DHBC syntheses, see [147]).

It is possible to use DHBCs as a tool to break down the three-dimensional symmetry of the interaction between nanoscale objects into dimension-specific communication [283]. Recent advances have demonstrated the possibility to self-assemble complex superstructures through the selective adsorption of functional polymers onto specific inorganic surfaces, thus inducing mesoscale self-assembly, cooperative transformation, and reorganization of hybrid inorganic/organic building blocks [40]. It has been shown that DHBCs [147, 284, 285] can exert a strong influence on the morphology and/or crystalline structure of inorganic particles such as CaCO$_3$ [61, 282, 286–292], PbCO$_3$, CdCO$_3$, MnCO$_3$ [61], calcium phosphate [108], barium sulfate [293–296], barium chromate [110, 297], barium titanate [298], calcium oxalate dihydrate [299], zinc oxide [288, 300–303], cadmium tungstate [304], chiral organic crystals [305], as a template for silica formation [306], and even on the control of the structure of ice and water [307]. These results show that not only the shape control of primary nanocrystals is achieved, but also that complicated superstructures can be produced through the mesoscale self-assembly of the "programmed" primary units.

All the above advantages of DHBC design also apply to the corresponding graft copolymers (DHGCs) with the further advantage of the easier synthetic accessability of graft copolymers. The main applications of DHBCs and DHGCs as crystallization additives will be described, ordered by typical DHBC applications. For a more extensive treatment of DHBC synthesis and applications, especially as an additive for crystallization control, the reader is referred to the specialized literature [57, 146, 147].

3.2.4.1
Formation and Stabilization of Nanocrystals

Various DHBCs have successfully been used as a stabilizer for the purely steric or combined steric/electrosteric stabilization of ceramic precursor particles like Al_2O_3 [308, 309], $BaTiO_3$ [310–314] colloidal silica [315–320] alumina coated TiO_2 [321], CeO_2 [322], α-Fe_2O_3 (hematite) [281, 323–325], and various pigments [327] in aqueous solution. The solid amount can be stabilized to extremely high concentrations (up to 80 wt % [327]) showing the stabilization capabilities of optimized DHBCs.

For example, the stabilization mechanism for α-Fe_2O_3 (hematite) particles by PMVE-b-PVOBA was investigated in detail for various block lengths addressing the adsorption isotherms, adsorbed layer thickness, and ζ-potentials [323, 324]. While PVME homopolymer did not absorb and PVOBA adsorbed as a polyelectrolyte at the applied pH, the highest polymer adsorption was observed for the block copolymer. The ζ-potential of the particles decreased with increasing PMVE block length due to the formation of a layer of neutral polymer providing steric stabilization [324]. These results could be confirmed by the stability of the dispersions stabilized with DHBCs with long PMVE blocks upon salt addition and indicate the role of the block copolymer architecture as outlined in Fig. 20 for a successful steric particle stabilization. Similar results were recently described for the stabilization of ACC by PEO-b-PMAA, which may be a useful building material for the construction of complex $CaCO_3$ morphologies [281]. Extremely small high quality CeO_2 nanoparticles of only 2 nm could be synthesized and stabilized with PEO-b-PMAA in a hydrothermal reaction [322]. This shows that the DHBC concept can also be applied to syntheses at elevated temperatures and that nanoparticles can be obtained with a very small size, which makes them attractive for the formation of homogeneous films or coatings with high surface area. If the nanoparticles are not directly accessible to DHBC adsorption due to equal charge and the resulting repulsion, the layer-by-layer assembly technique of countercharged polyelectrolytes can be successfully applied. This was demonstrated for the coating of maghemite nanoparticles by PEI and subsequent adsorption of a PEO-b-PGlu DHBC for excellent steric particle stabilization even at high salt concentrations [326]. These particles can be used for magnetic resonance imaging applications.

Recently, DHBCs have been used as a good stabilizer for the in-situ formation of various metal nanocolloids and semiconductor nanocrystals such as Pd, Pt [328–330], Au [280, 328–330], Ag [331], CdS [332], and lanthanum hydroxide [333]. PAA-b-PAM and PAA-b-PHEA were used as stabilizer for the formation of hairy needle-like colloidal lanthanum hydroxide through the complexation of lanthanum ions in water and subsequent micellization and reaction [333]. The polyacrylate blocks induced the formation of star-shaped micelles stabilized by the PAM or PHEA blocks. The size of the sterically stabilized colloids was controlled by simply adjusting the polymer-to-metal ratio, a very easy and versatile synthesis strategy for stable colloids in aqueous environment [333]. The concept of induced micellization of anionic DHBCs by cations was also applied in a systematic study of the direct synthesis of highly stable metal hydrous oxide colloids of Al^{3+}, La^{3+}, Ni^{2+}, Zn^{2+}, Ca^{2+}, or Cu^{2+} via hydrolysis and inorganic polycondensation in the micelle core [334, 335]. The Al^{3+} colloids were characterized in detail by TEM [336], and the intermediate species in the hydrolysis process by SANS, DLS, and cryo-TEM [337].

The formation of various metal colloids in the presence of DHBCs through either rapid reduction by addition of foreign reducing agents, or self-reduction in the presence of the DHBCs itself, can be an elegant way towards stabilized noble metal colloids in water [280, 328–331]. Through the DHBC functional block interaction with the cationic metal ions, micellization can be induced, where the micelle core acts as the reaction environment for metal salt reduction [328–330]. Such metal-filled micelles (Pd in PEO-b-P2VP) were successfully applied in water/isopropanol mixtures at different pH for a selective catalytic dehydrolinalol hydrogenation, showing the potential of aqueous DHBC/catalyst micelles [338].

Another systematic study showed that both the particle size and shape of Au could also be controlled by variation of the functional group pattern of the DHBCs and that the plasmon absorption band can be fine tuned depending on the type of polymer and the irradiation power [280]. PEO-b-PEI has been used as effective stabilizer for the solution synthesis of high-quality CdS nanoparticles with a size of 2–4 nm and good resistance against oxidation in water and methanol [332]. Here, the CdS particle size could be tuned by a simple polymer concentration variation, in analogy to the lanthanum hydroxide [333] or ACC [281] described above. This synthesis is easy and is performed in aqueous environment, in contrast to most of the toxic reaction environments for this material described in the literature. A comparison of a linear, branched, and dendritic functional DHBC block revealed that the branched block had the best stabilization capability, which was attributed to the high local functional group density combined with a certain molecular flexibility compared to the linear or dendritic counterparts [332]. The above examples imply that DHBCs can be applied to many more mineral or metal systems than have been described so far.

3.2.4.2
Stabilization of Specific Crystal Planes

The stabilization of specific crystal planes by face-selective polymer adsorption is a powerful tool for modifying the crystal morphology and for deriving low dimensional crystal morphologies like plates or fibers. Recent reports show that the DHBCs can also be used for stabilization of specific planes of some crystals for their oriented growth, e.g., Au [280], ZnO [300–302], calcium oxalate [299], PbCO$_3$ [291], and BaSO$_4$ [295].

A recent study shows that Au nanoparticles with shapes like triangles, truncated triangles, or hexagons, and a small fraction of smaller nearly spherical nanoparticles, can be obtained in the presence of a PEO-*b*-1,4,7,10,13,16-hexaazacyclooctadecan (hexacyclen) ethyleneimine macrocycle. The typical image in Fig. 21 shows the very thin and thus electron transparent triangles, truncated triangular nanoprisms, and the hexagons with high crystallinity as confirmed by the selected area electron diffraction pattern.

The results suggested that this polymer can well stabilize the (111) faces of cubic Au nanoparticles, leading to the preferential orientation and growth of other faces due to the selective absorption effects of the polymer. This results in the formation of the observed very thin plates. The unexpected, preferential, selective and strong polymer adsorption to the (111) faces can be understood on the basis of molecular modeling results, which show a good geometrical match of the interacting nitrogens in the hexacyclen part to the Au hexagons on the (111) face so that a high polymer surface coverage and thus extremely retarded growth of this face can be expected (Fig. 21b). The

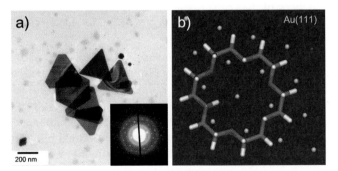

Fig. 21 a TEM image and electron diffraction pattern of Au nanoparticles synthesized by self-reduction of 10^{-4} M HAuCl$_4$ solution in the presence of the block copolymer PEO-*b*-1,4,7,10,13,16-hexaazacyclooctadecan (hexacyclen) EI macrocycle. **b** Molecular modeling of the Au (111) surface and a hexacyclen molecule in vacuum, which show the perfect match of this molecule to the hexagonal atom arrangement on Au (111). *Yellow* Au; *blue* N; *gray* C; *white* H. Figure drawn to scale. Reprinted from [280] with permission of American Scientific Publishers

span distance of the neighboring $-NH_2$ is 2.94 Å, which matches quite well with the distance of the neighboring Au atoms within the (111) face, as shown in Fig. 21b.

However, these considerations cannot be generalized. Application of the same DHBC additive for $CaCO_3$ crystallization did not result in crystal morphologies, which could be predicted from computer calculations of the additive adsorption onto different crystal faces [339]. Although interesting morphologies in the form of stacked pancake structures, discs with rough surfaces, or spheres were obtained, a conclusion of this study was that it is not only epitaxial match that drives polymer adsorption, but also charge/ion surface density, particle stabilization, and the time for polymer rearrangement. This polymer rearrangement to more favorable faces leads to adaptation of the thermodynamic crystal morphology, which becomes predictable again by calculation of the face-selective additive interactions [339].

Precipitation of ZnO under mild conditions usually produced elongated hexagonal prismatic crystal shapes with wide diameter range. Addition of PEO-*b*-PMAA resulted in a strong decrease of the particle size distribution width and decreasing crystal lengths, suggesting a good stabilization and selective adsorption effects of the DHBCs for a retardation of crystal growth perpendicular to (001) faces [300]. More rounded particles with a central grain boundary and a narrow size distribution were also produced in the presence of PEO-*b*-PMAA [301–303]. With the more strongly acidic PEO-*b*-PSS copolymers, a lamellar intermediate precipitate was found at the beginning, then this structure eventually dissolved and hexagonal prismatic crystals were produced. A striking effect was observed when a subsequent growth process produced uniform "stack of pancake"-shaped ZnO crystals. The formation of a metastable intermediate and the release of the polymer from the side faces of the core crystals, as well as the selective adsorption on the (001) ZnO faces, display a collective effect in the determination of such "stack of pancakes" morphology [303].

PEO-*b*-PMAA was also found to favor the morphological transition of COD crystals from the default tetragonal bipyramids dominated by the (101) faces to elongated tetragonal prisms dominated by the (100) faces, which could be due to the preferential adsorption of the polymer on the (100) face of COD crystals. With increasing polymer concentration, the morphology of the obtained COD crystals gradually changed from tetragonal bipyramids dominated by the (101) faces to rodlike tetragonal prisms dominated by the (100) faces, which is a morphology adopted by some plant COD crystals but not previously obtained in vitro [299].

Recent work shows that DHBCs can also be used to stabilize a single crystal face of $PbCO_3$ crystals [61]. In the presence of PEO-*b*-PMAA, disc-like nanoplatelets are formed as the primary short-term structure, in contrast to the poorly defined default star-like $PbCO_3$ structure under the chosen conditions. With increasing time, these nanoplatelets tend to branch or stack

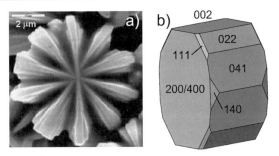

Fig. 22 a SEM micrograph of BaSO$_4$ single crystals with a symmetry-forbidden tenfold symmetry precipitated in the presence of PEO-b-PEI-SO$_3$H at pH 5. **b** The morphology of a primary crystal with ten side faces. Reprinted from [295] with permission of the American Chemical Society

together to form complex structures. When a new kind of DHBC, a stiff polymer PEO-b-[(2-[4-dihydroxy phosphoryl]-2-oxabutyl) acrylate ethyl ester] was used, flat uniaxially elongated quasi-hexagonal and single crystalline nanoplates with a smooth surface were obtained by selective polymer adsorption onto (001).

A striking example of complex morphologies that can be obtained by selective polymer adsorption showed that flower-like BaSO$_4$ particles with a forbidden tenfold symmetry can be produced by a multistep growth process in the presence of a sulfonated PEO-b-PEI, as shown in Fig. 22a [295].

For the formation of these particles, an elongated primary nanoparticle formed by selective polymer adsorption was proposed which has ten side faces (Fig. 22b, $4 \times (022)$, $4 \times (041)$ and $2 \times (002)$ family) [295]. This particle serves as a seed for the heterogeneous nucleation of ten single crystalline petals with the polymer selectively adsorbing onto (200), forming the petals with exposed (200), which are overgrown in a later stage. Thus a structure forms that is single crystalline (the petals) but which does not violate the laws of crystal symmetry due to the primary particle with ten side faces. It is remarkable that similar flower-like crystals, although not exclusively representing this unusual structure, could be obtained with a linear DNA template, which possibly acted as the template for the formation of a linear primary crystal [214]. These examples imply that symmetry-forbidden crystal structures are possible in a polymer-directed multistep crystallization process.

3.2.4.3
Control of Crystal Modification and Stabilization of Kinetic Phases

DHBCs have also been found to exert remarkable effects on the control of metastable crystal modifications and stabilization of kinetic phases. One example is a flexible DHBC containing multiple strongly chelating EDTA functional groups [286]. The interesting fact here is not that a vaterite sphere consisting of nanocrystallites was formed, as this is common for the early stages of $CaCO_3$ crystallization, but that the vaterite modification was stabilized for at least 1 year under conditions where a transformation of the kinetically favored vaterite modification into the thermodynamically stable calcite would occur within 80 h [340]. The remarkable stabilization of a metastable crystal modification suggests that the functional pattern of the flexible PEDTA block adopted that of the initially formed vaterite surface and that the crystal was fixed at the surface by the adsorbed polymer. Hence, a transformation into a thermodynamically stable crystal modification was not favored anymore because too many electrostatic bonds between Ca^{2+} and the polymer would have to be opened for restructuring. Similarly, prolonged stabilization of vaterite was observed with DHBCs with an acidic amino acid [341] or phosphoric acid functional group [292].

The stabilization of usually metastable crystal modifications hints at the potential DHBCs may have for crystal design purposes besides the morphological control, as many crystals of technical importance are made at high temperature to obtain the desired crystal modification. The same may be achieved by a polymer in water, via a kinetic metastable phase, if the functional block has the appropriate functional pattern, which requires a modular synthesis strategy towards DHBCs.

3.2.4.4
Chiral Discrimination and Morphogenesis of Organic Crystals

A series of chiral block copolymers were synthesized on the basis of a PEO-b-PEI$_{branched}$ copolymer with a variety of chiral functional groups attached to the PEI block as an additive to achieve racemate separation upon crystallization of racemic compounds [305]. In the crystallization of calcium tartrate tetrahydrate (CaT), it was found that appropriate chiral DHBCs can slow down the formation of the thermodynamically most stable racemic crystals as well as the formation of one of the pure enantiomeric crystals. Therefore, chiral separation by crystallization occurs even when racemic crystals are thermodynamically favored. However, the chiral separation was only observed (though still with low chiral purity) at the early stages of the crystallization reaction. At longer crystallization times the enantiomeric separation decreased again in favor of the racemate crystal formation. This indicates that a promising way for racemate separation with a chiral polymer additive is the separation of the crystals at early reaction stages so that an enrichment of one enantiomer appears possible via subsequent crystallization steps.

The presence of DHBCs can also modify the crystal morphology and create unusual morphologies of higher complexity reflecting the polymer–crystal interaction [305]. The basic principles of polymer-controlled crystal morphogenesis are also adaptable to organic crystals, which are promising for the investigation of polymeric crystallization control as they have a lower lattice energy than their inorganic counterparts. Furthermore, they are crystals built up from molecules, thus allowing for additional control parameters like inherent dipole fields in a molecule/crystal and chirality.

The most interesting observation was made when an amino acid was crystallized with one of the above-mentioned chiral DHBCs. The formation of facetted superstructures composed of highly aligned nanoparticle building units could be observed for DL-alanine, which was the first description of the mesocrystal concept [123].

This DL-alanine mesocrystal is shown in Fig. 23. The mesocrystal morphology (a) is modified with respect to the single crystalline needle obtained without polymer (b) and clearly shows the nanoparticle building units (c). The XRD scattering pattern is, however, very similar to that of the single crystal but in this case with a slight angular distortion (d and e). Dipolar fields were suggested for the almost perfect alignment of the nanocrystalline subunits because DL-alanine has a carboxy- and an amino-terminated end along the c-axis. Selective adsorption of the polyanionic DHBC to the amino-terminated end would lead to platelets, which then stack together to the mesocrystal with the platelet fine structure being visible in Fig. 23c.

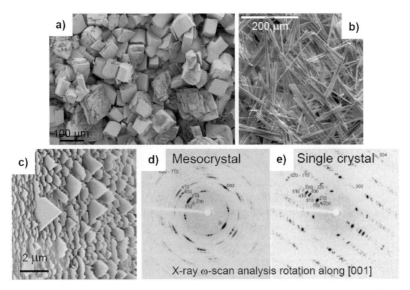

Fig. 23 DL-Alanine mesocrystals formed in presence of a chiral double hydrophilic block copolymer. **a** SEM micrograph showing the facetted mesocrystals, **b** DL-Alanine single crystal control, **c** Higher resolution SEM of a mesocrystal surface showing the nanoparticle building units, **d** XRD pattern of a mesocrystal, and **e** of a single crystal. The figures were reproduced from [123] with kind permission of Wiley

3.2.4.5
Mesoscopic Transformation and Controlled Higher Order Self-Assembly of Nanoparticles

Mesoscopic transformation [40] and controlled higher order self-assembly of nanoparticles are to date the most interesting morphogenesis mechanisms employing DHBCs, as they yield complex and often hierarchical structures by programmed self-assembly. In an ideal case, the self-assembly can even be encoded already in the block copolymer design to determine the crystal face for selective polymer adsorption so that the nanoparticle building units can assemble in a defined and programmed way. Nevertheless, the morphogenesis mechanisms are at present not yet completely understood.

The isomorphic $BaSO_4$ and $BaCrO_4$ systems are often applied as mineral model systems as they only crystallize in the barite and hashemite modifications, respectively. The platelet morphology in case of $BaSO_4$ remains the same over the entire pH range. Thus, $BaSO_4$ is very well suited to investigate the role of polymeric additives on crystal morphology, as the shape variation caused by different crystal modifications can be excluded here. Figure 24 gives an overview of the large variety of accessible single- and polycrystalline morphologies that can be obtained with DHBCs as additive.

Figure 24 illustrates the strong influence of the functional group pattern on an identical polymer backbone on the final crystal morphology [293, 294]. Without additives, rectangular $BaSO_4$ plates were formed. With the aid of different DHBCs, flower-like single crystalline structures, peanut shaped

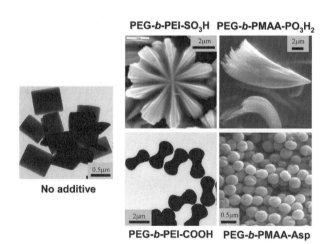

Fig. 24 Morphology control of inorganic $BaSO_4$ crystals mineralized at pH 5 in the presence of various DHBCs. Reprinted in part from [293] with permission of Wiley. Parts of the figure are reproduced from [294, 295] with permission of the American Chemical Society

polycrystalline particles monodisperse in size and shape, and polycrystalline spherical particles with notches were formed [293, 294].

It is noteworthy that the same polymer backbone (for example PEO-b-PEI in Fig. 24) can not only lead to entirely different crystal morphologies as mentioned above, but also to a single crystal in the case of the sulfonate functions, or to polycrystalline particles in the case of the weaker acidic carboxyl functions. Thus, it is evident that different morphogenesis mechanisms are active for the different DHBCs and the important role of the interacting functional polymer block is highlighted.

Crystallization control of the isomorphic $BaCrO_4$ by DHBCs is possible in a similar way [110]. It is interesting to note that all DHBCs based on a bare cationic PEI binding site give comparatively little control over the oriented growth of the crystals under the applied acidic conditions at pH 5, underlining that the growing crystals prefer to interact with negatively charged groups instead of the positively charged PEI block, a finding that is also valid for a number of Ca-based biomineral systems.

A block copolymer with carboxyl and phosphonated groups, PEO-b-PMAA-PO_3H_2, leads to the formation of single crystalline fiber bundles by oriented attachment (Fig. 24) similar to the bundles displayed in Fig. 14. However, the block copolymer is a far better particle stabilizer than polyacrylate so that a larger number of nanoparticle building units is generated and available for oriented attachment. As a result, the voids, which can be observed in the fiber bundles shown in Fig. 14, are not observed anymore for the structure obtained in presence of the DHBC. All other features of the bundles, like self-similarity or flat growth faces, are the same. Figure 25 shows an image of fibrous self-organized superstructures with sharp edges composed of densely packed, highly ordered, predominantly parallel nanofibers of $BaCrO_4$, which can be obtained with the phosphonated DHBC in analogy to $BaSO_4$ [110]. A selective area electron diffraction pattern taken from such an oriented planar bundle (as shown in Fig. 25) confirms that the whole structure scatters as a well-crystallized single crystal where scattering is along the [001] direction and the fibers are elongated along (210), suggesting a two-dimensional order of the nanofibers.

The atomic surface structure modeling data for the surface cleavage of the hashemite crystal shows that the faces ($1\bar{2}2$), ($1\bar{2}1$), ($1\bar{2}0$), and ($\bar{1}20$), which are parallel to the (210) axis, contain slightly elevated barium ions. This indicates that the negatively charged $-PO_3H_2$, $-COOH$ groups of PEO-b-PMAA-PO_3H_2 can preferentially adsorb onto these faces by electrostatic attraction and block these faces from further growth. In contrast, the surface cleavage of the (210) face shows no attackable barium ions on the surface (red lines in Fig. 25), suggesting that the negatively charged functional polymer group does not favorably adsorb on this face, making the (210) face a high energy face and thus favorable for oriented attachment, resulting in the nanoparticle fusion to defect-free fibers along (210).

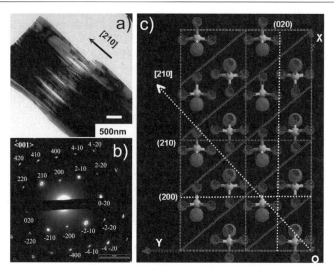

Fig. 25 a TEM image of highly ordered BaCrO$_4$ nanofibers obtained in the presence of PEO-*b*-PMAA-PO$_3$H$_2$, pH 5. **b** Electron diffraction pattern taken along the [001] zone, showing that the fiber bundles are well-crystallized single crystals and elongated along (210). **c** Plot of the BaCrO$_4$ crystal structure as calculated with the Cerius^2 software viewed perpendicular to the (210) axis (indicated by *white dashed arrow*). *Blue* = Ba, *red* = O, *light brown* = C. The *red lines* are (210) faces. **a,b** reprinted from [110] and **c** from [297] with permission of Wiley

The initially formed particles are amorphous with sizes of up to 20 nm and can aggregate to larger clusters. Evidently, this state of matter is the typical starting point for all types of highly inhibited reactions. The very low solubility product of barium chromate ($K_{sp} = 1.17 \times 10^{-10}$) shows that the superstructures do not really grow from a supersaturated ion solution but by aggregation/transformation of the primary clusters in agreement with other literature reports (as reviewed in [40]). This is in perfect agreement with earlier findings that in an ion solution with concentrations far above the saturation level, amorphous clusters are formed first, which then produce the crystalline nuclei at a later stage [243, 342]. The fibrous structures always grow from a single starting point, and from an aggregate of amorphous precursor particles, implying that they are nucleated on the glass wall or other substrate such as TEM grids, which obviously provide the necessary heterogeneous nucleation sites. In contrast, only spherical particles were obtained when the same reaction was done in polypropylene (PP) bottles [111].

The growth front of the fiber bundles is always very smooth, suggesting the homogeneous joint growth of all single nanofilaments with the ability to cure occurring defects in line with the earlier findings for BaSO$_4$ [109]. The opening angle of the cones is always rather similar, but depends on temperature, phosphonation degree of the polymer, and the polymer concentration [110].

Control experiments show that the higher the temperature, the more linear the structures become until the fibre structure is completely lost at 100 °C, where spherical or peanut-shaped particles were obtained. This could be due an increasing hydrophobicity of the PEO block at elevated temperatures, and increased particle mobility favoring uncontrolled nanoparticle aggregation over oriented attachment at lower temperatures. In addition, the concentration of the reactants, the ratio of cations and anions, pH, and the copolymer concentration all have a strong influence on the morphogenesis of $BaCrO_4$ because all these parameters determine the surface energies of the various nanocrystal faces [110, 297].

The whole suggested mechanism relies on the absence of flow, which would disturb the directed aggregation. Indeed, no fiber bundles or cones were obtained when the solution was stirred continuously at room temperature after mixing the reactants. Instead, only irregular and nearly spherical particles were obtained. This is again in agreement with a recently published mechanism where the fiber formation is due to the directed self-assembly of the primary particles [109, 111], which is suppressed by stirring.

It is interesting to note that a mixture of DHBCs, each of which result in defined $BaSO_4$ superstructures like spheres or slightly elongated particles (with a PEO-b-PMAA additive, Fig. 26c) and fibrous cones (with its partially phosphonated derivative, Fig. 26a) can result in cumulative or cooperative morphogenesis mechanisms, depending on the DHBC mixing ratio and the total additive concentration [333]. For example, shuttlecock morphologies

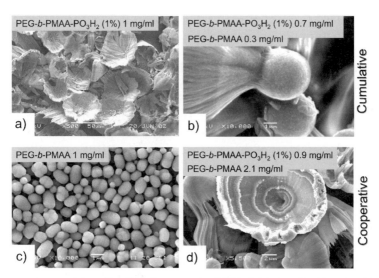

Fig. 26 $BaSO_4$ precipitated under the control of PEO-b-PMAA (**c**), its partially phosphonated derivative (**a**), and their mixtures at different total polymer concentration and mixing ratio (**b**), (**d**). Reproduced from [343] with permission of the Royal Society of Chemistry

could be obtained by secondary fiber cone nucleation onto initially precipitated spherical microparticles in a cumulative mechanism and in a predictable manner, as these structures are a superposition of a sphere and a cone or a fibrous cone nucleated on an initially formed sphere (Fig. 26b). This morphology is predictable and the extent of fibrous cone formation on the spherical template can be tuned by the DHBC mixing ratio. It should be noted that the sphere itself is already a polycrystalline aggregate controlled by PEO-b-PMAA, whereas the fibrous cone is controlled by the phosphonated derivative, which is a strong crystallization inhibitor so that the cone nucleation happens as heterogeneous nucleation on the sphere substrate.

However, if the total polymer concentration is increased, the entire nucleation becomes inhibited and a cooperative mechanism results in new complex fiber-related morphologies, which are not predictable from the crystal morphologies triggered by the individual DHBCs and strongly depend on the mixing ratio of the two functional DHBCs [343] (Fig. 26d).

These results show that the range of obtainable morphologies can be significantly enhanced by mixing of functional DHBCs. Another lesson that can be learned from this example is that even very simple mixtures of model polymers (here a DHBC and its phosphonated derivative) can result in complex morphologies via a so far unknown mechanism. This gives a taste of how difficult it is to understand the often synergetic actions of the multicomponent biopolymer mixtures, which control biomineralization events as a soluble matrix.

Calcium carbonate ($CaCO_3$) has been widely used as a model system for studying the bio-inspired mineralization process due to its abundance in nature and also its important industrial applications in paints, plastics, rubber, and paper [344]. Bio-inspired synthesis of $CaCO_3$ crystals in the presence of organic templates and/or additives has been intensively investigated and recent reviews show the wealth of results obtained in the last few years [44, 45].

$CaCO_3$ shows a broader morphological variety than $BaSO_4$ when crystallized in the presence of DHBCs. This is partly a reflection of the fact that $CaCO_3$ already has three different anhydrous polymorphs with different morphologies. Most of the reported studies on DHBC-controlled $CaCO_3$ crystallization deal with self-assembled structures whereas mesocrystal formation, oriented attachment, or PILP mechanisms have not yet been reported with DHBC additives. However, the formation of amorphous precursor particles is omnipresent in the DHBC-controlled crystallization of $CaCO_3$.

PEO-b-PMAA is probably the most extensively studied block copolymer so far. Its application was first independently reported in [286] and [287], although the reported structures were different (dumbbells and spheres in [286] and elongated overgrown calcite rhombohedra in [287]). Later on, a unifying overall scenario for this particular polymer–mineral system was evaluated, which allowed for an empirical morphology prediction if the pH as well as the polymer/mineral concentration ratio were known [282]. This result shows that already one particular polymer–mineral system can be tuned through a variety of particle aggregate morphologies, which are predominantly nanocrystal superstructures with complex morphologies, and that a uniform growth mechanism for several of these superstructures can be assumed. One such mechanism, which was identified for $CaCO_3$ in this particular system, is the rod–dumbbell–sphere morphogenesis scenario, in formal analogy to the results obtained for fluoroapatite in a gelatin gel [345–347].

The morphology evolution process of $CaCO_3$ particles under control of DHBCs suggests that the final stage of spherules results from a complex growth mechanism, as first reported in detail for the growth of fluoroapatite in gelatin gels by Busch and Kniep [345, 346]. In the fluoroapatite case, rod-like particles are formed first, which can grow at their ends to result in dumbbell-like particles due to a dendritic growth [345–347]. These dumbbells grow further into spherical particles. Such fluoroapatite morphologies are intuitively consistent with a particle growth along electric dipole field lines, although this mechanism has not yet been proven [346, 348]. This morphology evolution process is qualitatively in agreement with the dumbbells and spheres observed for $CaCO_3$ mineralized in the presence of DHBCs. Here, however, the complex morphologies appear to result from a nanoparticle aggregation mechanism, as WAXS data evaluated by the Scherrer equation reveal small primary nanoparticle building units with sizes around 20 nm [286]. A time-resolved electron microscopy study indeed revealed that dumbbells are the precursors to the much bigger spherical particles and that amorphous precursor nanoparticle aggregates are present, which most likely act as a material depot [282].

Nevertheless, the rod-to-dumbbell-to-sphere transition seems to be a general crystal growth phenomenon and was observed for several other carbonate systems in the presence of DHBC additives ($CaCO_3$, $BaCO_3$, $MnCO_3$, $CdCO_3$) [61] as well as in rare cases even without additives. Thus it appears that the polymer additive does not only play a structure-directing role as the above discussion may imply, but also a controlling role for the release of building material consisting of amorphous particles.

An interesting morphogenesis strategy, which can be realized with DHBC additives, is the generation of sacrificial precursor particles of a metastable crystal polymorph. It is well documented that DHBCs can stabilize thermodynamically metastable polymorphs like vaterite for extended periods of time, even in aqueous solution [113]. Therefore, it can be a rewarding strategy to apply a DHBC, which is able to control the morphology of a metastable polymorph but is only able to stabilize this polymorph for a time not much extending that required for generation of the sacrificial precursor particle.

An example was reported for hollow spheres composed of calcite rombohedra, which where grown on a sacrificial spherical vaterite template particle grown in the presence of PEO-b-PMAA at an acidic starting pH [61]. Initially, amorphous nanoparticles were formed and stabilized by the DHBC. Subsequently, vaterite nanoparticles were formed, which aggregated to spheres. These vaterite spheres were the template particles for the nucleation of calcite via a dissolution–recrystallization process (Fig. 27). Similar scenarios can be imagined for a large variety of metastable or unstable phases including liquid, amorphous, kinetically labile or metastable polymorph phases.

It is remarkable that hollow calcite spheres, but with less expressed surface rhombohedra, have been reported when the same block copolymer was used together with the SDS surfactant [245]. These rhombohedral surface structures are very similar to those reported by Tremel et al. when Au nanocolloids

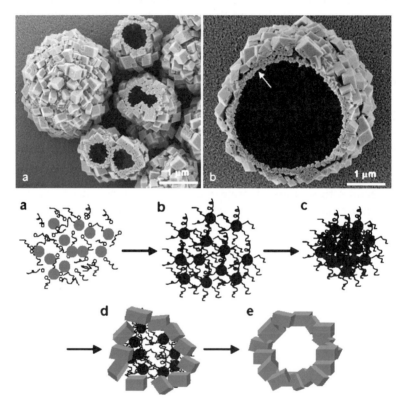

Fig. 27 *Top*: SEM images of $CaCO_3$ particles grown on a glass slip in the early reaction stage, PEO-*b*-PMAA, $[CaCl_2]$ = 10 mM, 1 g L^{-1}, 5 h. **a** $CaCO_3$ particles with either spherical or hollow structures. **b** Zoom showing the calcite rhombohedral subunits grown on the surface of the hollow structure and the inner part consisting of tiny primary nanocrystals with a grain size of about 320 nm as indicated by the *arrow* (sacrificial vaterite template). *Bottom*: Proposed formation mechanism of the calcite hollow spheres: *a* polymer-stabilized amorphous nanoparticles; *b* Formation of spherical vaterite precursors; *c* aggregation of the vaterite nanoparticles; *d* vaterite–calcite transformation starting on the outer sphere of the particles; *e* formation of calcite hollow spheres under consumption of the sacrificial vaterite precursors. Reproduced in part from [61] with permission of the American Chemical Society

were used as nuclei for crystallization of calcite spherules with sizes of about 10–15 μm [349, 350].

Spherical vaterite microparticles were also found if acidic polypeptide blocks like PGlu were used as functional blocks in a PEO-based DHBC [341, 351]. These structures are again formed from amorphous precursor nanoparticles (50 nm), which are remarkably uniform. However, in contrast to the calcite rod–dumbbell–sphere morphology transition that is observed for calcite with a PEO-b-PMAA additive, in the vaterite case, spherical particles are directly formed without any rod or dumbbell precursors, as shown in solution by dark field and phase contrast light microscopy. It is remarkable that the structure-directing polymers do not differ much. Both contain a PEO stabilizing block of similar length and a functional polycarboxylic acid block (PMAA$_7$ in the calcite case and PGlu$_{10}$ in the vaterite case). This indicates that subtle changes of the functional DHBC block can have a huge influence on the crystallization scenario. In addition, it could be shown that a random coil conformation of the polypeptide chain allowed for better crystallization control than the ordered α-helix conformation [351], which is in agreement with the findings in the literature about biomineralization proteins, which tend to adopt the unfolded structures for their interacting motifs [352].

The most advanced way to encode nanoparticle self-assembly is to selectively adsorb additives onto specific crystal surfaces and thus to program their self-assembly. Such a possibility approaches the ability of nature to transform soft structures, which can be encoded in genetic information and can thus be highly regulated into hard organic/inorganic replicas through the process of biomineralization.

One example of such a polymer-encoded self-assembly process was demonstrated for BaCO$_3$ crystallization in presence of a stiff phosphonated DHBC [353]. Here, the selective adsorption of the stiff DHBC onto the (110) witherite surfaces leads to the tectonic arrangement of the elongated orthorhombic BaCO$_3$ by programmed self-assembly resulting in remarkable helical structures from a non-chiral mineral crystal system and a racemic polymer (Fig. 28a). The amount of left- and right-handed helices was found to be similar [353].

This arrangement via coded self-assembly relies on two processes. The first is the existence of favorable (110) faces for the adsorption of the stiff DHBC, leading to a staggered arrangement of aggregating nanoparticles, being controlled in direction after the aggregation of the first three particles (Fig. 28d–g). The driving force is the loss of conformational entropy of the polymer if it would have to bend after aggregation of another nanoparticle building unit into an unfavorable position, and the face-selective polymer adsorption onto the (110) face, which exposes positive Ba^{2+} ions so that the polyanion is attracted. The result is a nanoparticle aggregation along a common growth direction, which is determined by the aggregation of the first three building units.

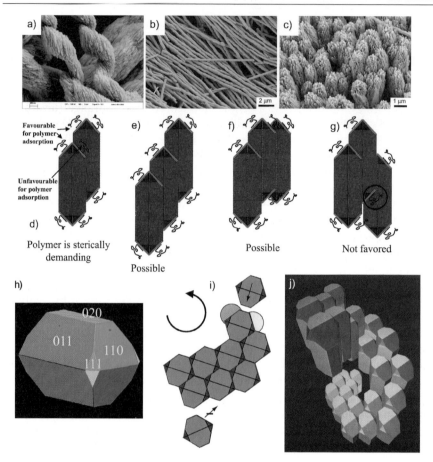

Fig. 28 a–c SEM micrographs of BaCO$_3$ nanoparticle superstructures obtained under control of a stiff phosphonated DHBC, [Ba^{2+}] = 10 mM. **a** Helices formed at starting pH 4, $c_{Polymer}$ = 1 g L^{-1}; **b** starting pH 5.6, $c_{Polymer}$ = 1 g L^{-1}; **c** starting pH 4, $c_{Polymer}$ = 2 g L^{-1}. The *colors* in **d–g** correspond to the indexed faces of the primary nanocrystal witherite building block in **h**. **d** Interparticle aggregation along the (011) and (020) faces makes the adsorption faces at the tip different, allowing the particles to differentiate between favorable and unfavorable arrangements. The (partially) staggered arrangement is preferred (**e,f** with **f** being much more improbable) over the probable but energetically excluded arrangement with unfavorable polymer adsorption sites (*circle* shown in **g**) and predetermines addition of subsequent particles. The result is directed aggregation of subsequent particles either upwards or downwards as defined by the first three aggregated particles. **i** Perpendicular view along the helix growth axis. The alternation of particle aggregation by (011) (*red lines*) and (020) (*black lines*) faces breaks the linear character and brings in helicity {only one (020) face sterically accessible in a favorable way}. The *yellow and orange spots* indicate different (011) faces for an attaching particle. The *orange spot* is more favorable for subsequent particle attachment and thus, the helix turn is continued. Overlay of processes **d–g** and **i** leads to the helical superstructure **j**, which even tolerates some mismatched particles. Please note that in Fig. 28, the polymer is not drawn as a stiff rod and also not to scale. Figure partly reproduced from [151] and [353] with permission of Nature Publishing Group and the Materials Research Society

On the other hand, in the perpendicular direction (Fig. 28i), a particle approaching an aggregate can also envisage favorable and unfavorable adsorption sites leading to a turn in the particle aggregate. If a particle aggregates along (020), a linear arrangement will result, whereas two possibilities exist for an approaching particle to aggregate along the (011) faces (Fig. 28i). These possibilities are indicated by a yellow (unfavorable) and orange (favorable) spot in Fig. 28i. The yellow position is unfavorable because the approaching particle has to be oriented along two faces, whereas for the orange position, the orientation along (011) is sufficient. Therefore, this will be the preferred adsorption site. This introduces a counterclockwise turn in the particle aggregate in the example in Fig. 28i.

Overlay of these two processes leads to the helix formation shown in Fig. 28j. The pitch of the helices can be controlled by influencing the counterplay between the two processes controlling self-assembly. For example, increasing the pH (Fig. 28b) favors the nucleation of nanoparticle building units and consequently the particle aggregation. Furthermore, the nanoparticle building units get less charged as the raised pH approaches the isoelectric point of $BaCO_3$ (10.0–10.5) so that particle repulsion is decreased and particle assembly favored.

On the other hand, increase of the polymer concentration leads to increased polymer adsorption and also to less favorable (110) faces so that no axial aggregation can occur anymore. The result is a back biting helix without pitch like that in Fig. 28c. However, the above extreme examples show that even the pitch of the tectonic helix arrangements can be controlled by a parameter as simple as pH or polymer concentration. This example shows the delicate balance necessary for programmed particle self-assembly and the possibilities of such approach, which can even create chirality from non-chiral/racemic compounds.

Besides programmed self assembly, synergy between polymer-controlled crystallization and crystallization-induced reordering of a polymer aggregate structure can create a feedback loop and result in highly complex organic/inorganic mesostructures. A striking example for such a mechanism was reported for calcium phosphate precipitated within an aggregate of poly(ethylene oxide)-*b*-alkylated poly(methacrylic acid) (PEO-*b*-PMAA-C_{12}) [108]. In this polymer, the functional carboxy groups are in the direct neighborhood of the hydrophobic alkane chains. This is special in the way that the three hydrophobic chains attached to the functional methacrylic acid block induce a weak aggregate formation in water, as evidenced by dynamic light scattering (DLS) and analytical ultracentrifugation (AUC). These loose aggregates sequester Ca^{2+} ions and thus serve as localized mineralization centers upon phosphate addition.

By controlling the pH of the Ca^{2+}-loaded polymer solution and thus the crystal polymorph (brushite and hydroxyapatite), delicate mesoskeletons of interconnected calcium phosphate nanofibers with star-like, neuron-like, and more complex nested form can be obtained as shown in Fig. 29 [108]. The filaments are very thin with only 2–3 nm diameter and they are oriented along the hydroxyapatite c-axis.

The complex structures result from a feedback loop between selective polymer adsorption on all faces parallel to the hydroxyapatite c-axis and the deformation of the polymer aggregate by the polymer adsorption onto the growing filaments. From Fig. 29d, it becomes clear that the negatively charged DHBC will adsorb to all faces that are parallel to the hydroxyapatite c-axis, as these faces can expose Ca^{2+} ions. The only face exposing negatively charged phosphate tetraeders is (001) so that the polymer adsorption to all faces except (001) leads to the filament formation with c-axis orientation.

The above example also shows that the three-dimensional structure of the crystallization additives prior to mineralization provides a further variable for the control of crystal morphogenesis by polymers, which even more closely resembles natural protein systems in the soluble matrix of biominerals.

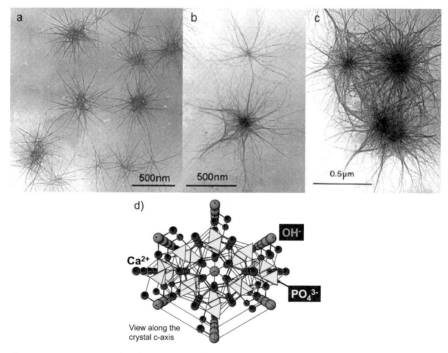

Fig. 29 TEM images of calcium phosphate block copolymer nested colloids. **a** Star-like form at an early stage at pH 4.5. **b** Later stage showing a complex central core, and very long and thin (2–3 nm) filaments. **c** Neuron-like tangles produced at pH 5. **d** View along the hydroxyapatite c-axis. Four unit cells (unit cell indicated by *red line*) are displayed. Horizontal Ca^{2+} distance is 2.2 nm and vertical Ca^{2+} distance is 1.7 nm, which fits the experimentally observed fiber diameter. Reproduced from [108] with permission of Wiley

3.2.4.6
Combination of DHBCs with External Templates

From the above examples, it is clear that DHBCs can promote the formation of complex crystal morphologies, often involving nanoscopic building units. However, it is known from biomineral examples that their formation takes place in confined reaction environments and in contact with external templates. Thus, it is interesting to investigate the influence of an external template on a DHBC-controlled crystallization system. In one reported example, the chosen template was as simple as CO_2 gas bubbles, which are generated by CO_2 evaporation from supersaturated $Ca(HCO_3)_2$ solutions (Kitano method). Here, the gas bubbles, which emerge in solution and float to the surface where they stay for a while, were used to structure DHBC-generated $CaCO_3$ nanoparticle building blocks to complex morphologies including polycrystalline flower, shuttlecock, and hollow half-sphere morphologies [292]. The morphologies could be tuned via variation of the solution surface tension. This was adjustable by the variation of the polymer phosphorylation degree. $CaCO_3$ nanoparticles were generated and temporarily stabilized by the DHBC until they aggregate as a ring around a CO_2 bubble at the solution surface. This ring closes by further particle aggregation and the structure sinks to the bottom of the reaction vessel and gets quenched from further growth as soon as it became too heavy to be held at the surface by the solution surface tension. Adjustment of the surface tension thus allows control of how complete the particle aggregate structures can grow. This shows a simple way to complex $CaCO_3$ morphologies via the concept of polymer-stabilized nanoparticle building units.

Similarly, a combination of the mineralization under control of low molecular weight polyelectrolytes and a foreign static template such as air bubbles has been explored for the generation of macroporous $BaCO_3$ spherulites [230], which is similar to that found in the microemulsion system reported by Mann et al. [354].

However, particles can also be used as templates instead of gas bubbles. In order to break the previous constraint of bundle formation in $BaSO_4/BaCrO_4$ crystallization controlled by a phosphonated DHBC [109, 110], a new and potentially facile stage of the morphological control of inorganic materials using a combination of crystal growth control by DHBCs and controlled nucleation was applied, resulting for the first time in the fabrication of uniform, separated $BaCrO_4$ single crystalline nanofibers with extremely high aspect ratios of > 5000. This was achieved by adding a minor amount of a cationic colloidal particle, thus enriching the DHBCs as well as Ba^{2+} in a confined area of space and reaching high local supersaturation and nucleation close to these spots [297]. The attachment and growth of a second fiber onto an already formed one was effectively hindered by adsorption of the colloidal seed particles to the sides of existing fibers, thus acting as sterical blockers.

If DHBCs are encoded in a crosslinked microgel architecture, they can be used as a template for crystallization. One advantage of such an approach is that the microgels can serve as a localized crystallization environment as they are able to sequester ions and thus act as ion sponges. A PMAA-b-PNIPAAM-b-PDEAEMA microgel with PMAA core was used for $CaCO_3$ crystallization [127]. The metastable aragonite polymorph was nucleated at ambient temperature and pressure and it could be shown that the applied microgels act as a highly selective nucleation and polymorph control agent, which is active at low concentrations down to only 0.1 ppm. The advantage is that, consequently, the nucleated aragonite particles are essentially free from organic contaminants and can be used as they are. The adapted aragonite morphology was that of a sheaf bundle, which was nucleated in the microgel particle with progressing growth of a ring of aragonite nanocrystals. Interestingly, the aragonite sheaf bundle superstructure is composed of a large number of individual nanocrystals, which are coated by an ACC layer only a few nanometers thick [127] as was also recently found in natural aragonite in nacre [72]. However, the formation of a large number of aragonite nanoparticles from a single template particle implies that the individual nanoparticles are connected by mineral bridges. This is a synthetic analog of the finding in biomineralization that a tiny amount of additive can influence a large amount of inorganic crystals.

3.2.4.7
DHBC–Surfactant Complexes as Crystallization Templates

The combination of crystallization control by DHBCs and self-organization of surfactants in aqueous environment can lead to remarkable new crystalline structures as also reported for similar but more simple polyelectrolyte/surfactant additives [355, 356]. Elegant featherlike $BaWO_4$ nanostructures were prepared under mild conditions in a multistep growth mechanism by combination of both catanionic reverse micelles (undecyl acid and decylamine) and the block copolymer PEO-b-PMAA itself [357]. Numerous, nearly parallel single crystalline barbs stand perpendicular on both sides of a polycrystalline central shaft, showing that special template effects can be achieved in a limited experimental window (Fig. 30).

These structures could be varied from star-like structures to a single shaft by a simple variation of the DHBC concentration, although the role of the DHBC in the generation of these complex structures initially remained unclear.

The morphogenesis of the feather-like structures in Fig. 30 was recently investigated in more detail. The same type of structures was generated with catanionic reverse micelles [358]. These results implied a two-stage morphogenesis mechanism, where first a $BaWO_4$ shaft is generated under the control of PEO-b-PMAA. Subsequently, [001] oriented nanobelts then grow perpendicular to the shaft under the control of the surfactant additive. This example

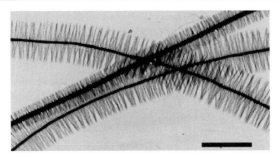

Fig. 30 TEM image of penniform BaWO$_4$ nanostructures obtained in the presence of 0.5 g L^{-1} PEO-b-PMAA after aging for 8 h in reverse micelles. *Scale bar* represents 5 μm. Reprinted in part from [357] with permission of the American Chemical Society

shows again that complex structures can be generated by mixtures of additives if a time axis can be utilized with the action of the two additives being separated in the time domain (see also Fig. 26 [343]).

In addition, hollow structures of calcite and disc-like hollow vaterite particles can be obtained by the cooperative template effects of the complex micelles formed by PEO-b-PMAA and SDS and remaining free DHBC as inhibitor in solution [245]. The cationic surfactant CTAB, which can complex to the anionic PMAA groups of the DHBC, on the other hand yielded unusual calcite pine-cone shaped particles. The concept of PEO-b-PMAA-SDS micelles as template was further extended to the production of hollow submicrometer-sized Ag spheres [359].

3.2.4.8
Fine-Tuning of the Nanostructure of Inorganic Nanocrystals by Adjusted Crystallization Conditions

In addition to the above reported superstructure formation under control of block copolymers, the DHBCs can also be used to fine-tune the nanostructure details of other inorganic nanocrystals. Very thin one-dimensional and two-dimensional CdWO$_4$ nanocrystals with controlled aspect ratios were conveniently fabricated at ambient temperature or by hydrothermal ripening under control of DHBCs [304]. Simple temperature changes could be applied to tune the particle morphologies between plates and rods.

3.2.4.9
DHBC-Controlled Crystallization in a Mixed Solvent

When the crystallization conditions of CaCO$_3$ in the presence of a DHBC are modified by applying water–alcohol solvent mixtures with varying solvent compositions the solvent quality changes for the block copolymer and

$CaCO_3$. Thus, further handles are available to tune the higher order assembly of $CaCO_3$ nanoparticle aggregates including well-defined elongated or spherical morphologies [360]. However, as worsening of the solvent quality for DHBCs is against the concept of well-soluble polymer additives, such strategy develops more towards externally triggered DHBC aggregation, which may be an interesting additional experimental variable as cation-induced micellization has shown. Further, as the mineral solubility product is simultaneously changed, interesting cooperative morphogenesis scenarios can be expected with DHBC additives, as the morphology and polymorph of $CaCO_3$ is already influenced by the composition of mixed water–alcohol solvents [361].

3.2.5
Double Hydrophilic Graft Copolymers (DHGCs)

Double hydrophilic graft copolymers also follow the concept of DHBCs and essentially behave similarly. They are potentially interesting because their synthesis is easier than that of the DHBCs and can be scaled up. Nevertheless, only a few examples have been reported so far for crystallization control in aqueous environments applying DHGCs [362]. Wegner et al. applied poly(ethylene oxides) connected to the carboxylic acid group of methacrylic acid to prepare poly(ethylene oxide) graft copolymers with methacrylic acid and/or vinylsulfonic acid by free radical polymerization [362]. These polymers were active in the control of ZnO crystallization and the ZnO morphology could be controlled by the vinylsulfonic acid content in the block copolymer under otherwise unchanged conditions, where vinylsulfonic acid was much more active than the weaker methacrylic acid [362].

Poly(ethylene oxide) graft copolymers with a polyacetal backbone resulting in graft copolymers with functional carboxy groups along the main chain were also reported [363]. These polymers were applied as additives in the crystallization of $CaCO_3$ and the influence of the variation of molecular parameters on the crystallization was systematically studied with respect to dispersion stability, particle size, and shape [364]. Variation of the block lengths of the hydrophilic and ceramophilic blocks resulted in altered calcite particle size, shape and dispersion stability, and aggregation behavior so that an optimum balance between both moieties in the graft copolymer was found to be essential for maximum effect on the $CaCO_3$ crystallization behavior [364]. This is in analogy to the considerations for DHBCs in Fig. 20.

4
Conclusion

In this overview, several strategies of bio-inspired crystallization with hydrophilic polymer additives have been presented. The most well known one is the strategy to influence a classical crystallization reaction, proceeding by addition of ions to crystal faces, by face-selective adsorption of the polymeric additive. The polymer adsorption will lower the interfacial energy of this face so that its growth velocity is decreased. This strategy, which relies on thermodynamic principles, would in principle even lead to predictable crystal morphologies according to Wulff's rule. This rule describes the equilibrium morphology of a crystal as the minimum energy solution of the sum of the product of interface area and energy [58]. However, this is only rarely the case, as additive-controlled crystallization mainly proceeds along kinetic crystallization pathways so that the prediction of the resulting morphology is almost impossible, even if the crystallization system, additive, and conditions are well known. However, face-selective polymer adsorption can still lead to remarkable effects as it can program a subsequent self-assembly process, which can even lead to chiral structures from achiral/racemic compounds [353]. Combination of a template and additive-mediated self-assembly, additive-controlled crystallization coupled with crystallization-mediated polymer aggregate rearrangement, and polymer or polymer–surfactant mixtures with both additives acting on a different time-scale can also lead to very complex structures. In general, the inspiration taken from biomineralization archetypes points towards the way of complex and often coupled additive mixtures, with components acting at different time-scales and in different localized reaction compartments. In addition, self-assembly of preformed nanoparticles is an important strategy towards complex morphologies.

Mimicking these natural strategies, even with very simplified synthetic systems, has already led to remarkable results concerning the generation of complex mineral morphologies and control over crystallization events. However, most of the crystallization mechanisms are still unknown due to the complexity of the system, which involves time-dependent structures with sizes spanning the entire colloidal level.

However, the strategy to copy natural biomineralization systems is just one side of the coin. Bio-inspired mineralization can also lead to the exploration of new crystallization pathways, which might also be of importance in biomineralization processes but have not yet been discovered in these systems due to their complexity, so that bio-inspired simplified model systems are clearly desirable. Mechanisms, which have been discovered in bio-inspired mineralization systems are "oriented attachment" as well as crystallization via amorphous or even liquid precursors and mesocrystal precursors. These mechanisms do not all proceed according to the textbook knowledge of crystallization so that they can be summarized under the term "non-classical crystallization". A common feature is that all of these mechanisms lead to the formation of a final single crystal via particle (or liquid droplet) precursors instead of the classical ion-by-ion building of a single crystal. These nanoparticle-based pathways imply that the definition of a single crystal formed by classical crystallization should be based on a single nucleation event, whereas non-classical nanoparticle-mediated crystallization is associated with multiple nucleation events.

While amorphous precursors were already found and discussed for biomineral systems as a sophisticated way towards minerals as they avoid high salt concentrations with their associated high osmotic pressures, the other mechanisms emerged from purely synthetic systems. Only very recently, the first evidence was reported that, for example, mesocrystals can be found in biominerals like aragonite platelets in nacre [114] or calcite nanocrystals in sea urchin spines [116]. Much has still to be done to understand the forces that control the perfect nanoparticle alignment in mesocrystals, as well as the exact building mechanism and the possibilities for manipulating these structures. However, the toolbox of crystallization is clearly extended now.

This means that bio-inspired mineralization can in turn also lead to inspiration for the exploration of biomineralization systems as the bio-inspired systems can be kept as simple as necessary to reveal the required detail in a complex mineralization process. Current developments undoubtedly emphasize that probably all inorganic crystals will be amenable to morphosynthetic control by use of either flexible molecule templates or suitable self-assembly mechanisms. These emerging new nature-inspired solution routes may open alternative pathways towards low dimensional nanocrystals and more complex superstructures. Further exploration in these areas should provide new possibilities for rationally designing various kinds of inorganic materials with ideal hierarchy and controllable length scale. These unique hierarchical materials of structural specialty and complexity, with a size range spanning from nanometers to micrometers, are expected to find potential applications in various fields. In addition, it is undoubtedly necessary to investigate the relationship between the structural specialty/complexity (shape, size, phase, dimensionality, hierarchy etc.) of the materials and their properties, which could result in novel applications as building blocks in various fields of materials science and other related fields. As a bottom line, it is clear that the textbook knowledge on crystallization needs to be extended and that a further understanding of alternative particle-mediated crystallization pathways, as an extended crystallization toolbox, will give insight into alternative future ways for the generation of advanced materials.

5
Current Trends and Outlook to the Future

In the last few years, the number of reports about the application of bio-inspired crystallization mediated by hydrophilic polymers has significantly increased. This is triggered by increased knowledge gain in three important key fields:

1. Biomineralization processes are increasingly well understood as a result of better analytical capabilities. Therefore, alternative crystallization scenarios, like that via amorphous precursor phases and very recently even via mesocrystals, have become visible. On the other hand, more and more biomineralization polymers (mainly proteins) have been separated, characterized, and sequenced so that the molecular knowledge base on biomineralization additives is also increasing. All these efforts already allow a view on the mechanisms of polymer-controlled crystallization, even though it is still fragmented.
2. Synthetic polymer chemistry has made significant advances towards controlled radical polymerization reactions and solid state peptide syntheses so that model polymers for biomineralization polymers have become available by easier synthesis procedures than the ionic polymerization routes.
3. The evidence for non-classical particle-mediated crystallization reactions has increased over the last few years. Oriented attachment, liquid precursors, as well as mesocrystal crystallization pathways have shown their potential to significantly expand the crystallization toolbox in materials chemistry. On the other hand, biomineralization processes can be seen from a new perspective.

A shift from the descriptive treatment of reaction products under various experimental crystallization conditions to an analytical consideration of the underlying mechanisms of the observed and often amazing controlled crystallization reactions can be stated. Also, a trend can be observed in that the bio-inspired model systems are increasingly being chosen to answer a specific question and to reveal mechanistic details, rather than the empirical approaches of the past, which were descriptive in nature but helped to set the field for more promising systematic investigations. Continuation of the present research effort on bio-inspired mineralization with hydrophilic polymer additives will possibly even reveal further non-classical crystallization pathways and will lead to an increased understanding in important areas of general scientific interest, for example:

1. Description of crystallization as an important phase transformation process from a more generalized viewpoint.
2. Controlled nanoparticle self-assembly over several length scales.
3. Transformation of morphological information encoded in soft organic matter to organic/inorganic hybrid systems.
4. Exploration of new material design strategies and generation of materials with improved chemical and physical properties. The approach from nature is adaptable to functional materials as well.

5. Treatment of biomineralization-related diseases like bone or teeth defects etc. with repair possibilities as well as implant design.
6. Nanotechnology in general as bio-inspired crystallization also concerns the spatially controlled deposition of functional crystalline units. How can self-assembly be triggered and controlled in complex multicomponent systems?
7. Biology can profit from an increased understanding of biomineralization processes to reveal the design strategies of nature's sculptures.
8. Physical chemistry and physics can profit from the emerging alternative particle-mediated crystallization pathways, which have in common that they are multistep processes that follow a kinetic pathway. New physical insights into the structuring of matter can be gained. For example, the reason why nanocrystals self-assemble in a three-dimensional crystallographically aligned manner in a mesocrystal is not yet revealed.

Reflection on the above points already indicates that an even further increased scientific interest into polymer-controlled crystallization can be expected. This field is multidisciplinary because it is of interest for multiple scientific disciplines. Bio-inspired mineralization is straight at the heart of several key development fields for future technological benefit. Therefore, it will be of importance that multidisciplinary teams of researchers share their ideas and results in this complex field.

Acknowledgements I thank the Max Planck Society, the Fonds of the German Chemical Industry, the "Bundesministerium für Bildung und Forschung (BMBF)" as well as the German Science Foundation (DFG) within the Special Research Program "Principles of Biomineralisation" for financial support of my ongoing research on bio-inspired mineralization. Nicole Gehrke is acknowledged for useful discussions and proofreading of this manuscript.

References

1. Dabbs DM, Aksay IA (2000) Annu Rev Phys Chem 51:601
2. Matijević E (1993) Chem Mater 5:412
3. Matijević E (1996) Curr Opin Colloid Interface Sci 1:176
4. Antonietti M, Göltner C (1997) Angew Chem Int Ed 36:910
5. Archibald DD, Mann S (1993) Nature 364:430
6. Yang H, Coombs N, Ozin GA (1997) Nature 386:692
7. Li M, Schnablegger H, Mann S (1999) Nature 402:393
8. Peng XG, Manna L, Yang WD, Wickham J, Scher E, Kadavanich A, Alivisatos AP (2000) Nature 404:59
9. Ahmadi TS, Wang ZL, Green TC, Henglein A, El-Sayed MA (1996) Science 272:1924
10. Gibson CP, Putzer K (1995) Science 267:1338
11. Pileni MP, Ninham BW, Gulik-Krzywicki T, Tanori J, Lisiecki I, Filankembo A (1999) Adv Mater 11:1358

12. Park SJ, Kim S, Lee S, Khim ZG, Char, Hyeon T (2000) J Am Chem Soc 122:8581
13. Adair JH, Suvaci E (2001) Curr Opin Colloid Inter Sci 5:160
14. Yu SH, Yang J, Qian YT (2004) Low dimensional nanocrystals. In: Nalwa HS (ed) Encyclopedia of nanoscience and nanotechnology. American Scientific Publishers
15. Lieber CM (1998) Solid State Commun 107:607
16. Hu JT, Odom TW, Lieber CM (1999) Acc Chem Res 32:435
17. Klein JD, Herrick RD, Palmer D, Sailor MJ, Brumlik CJ, Martin CR (1993) Chem Mater 5:902
18. Martin CR (1994) Science 266:1961
19. Routkevitch D, Bigioni T, Moskovits M, Xu JM (1996) J Phys Chem 100:14037
20. Duan XF, Lieber CM (2000) Adv Mater 12:298
21. Duan XF, Lieber CM (2000) J Am Chem Soc 122:188
22. Zhou Y, Yu SH, Wang CY, Li XG, Zhu YR, Chen ZY (1999) Adv Mater 11:850
23. Stein A, Keller SW, Mallouk TE (1993) Science 259:1558
24. Trentler TJ, Hickman KM, Goel SC, Viano AM, Gibbons PC, Buhro WE (1995) Science 270:1791
25. Yu SH, Yoshimura M (2002) Adv Mater 14:296
26. Estroff LA, Hamilton AD (2001) Chem Mater 13:3227
27. Niemeyer CM (2001) Angew Chem Int Ed 40:4128
28. Wood R, Grotzinger JP, Dickson JAD (2002) Science 296:2383
29. Mann S (1997) Chapter 5. In: Bruce DW, O'Hare D (eds) Inorganic materials. Wiley, p 256
30. Lowenstam HA, Weiner S (1989) On biomineralization. Oxford University Press, New York
31. Mann S, Webb J, Williams RJP (1989) Biomineralization. Wiley, Weinheim
32. Mann S (2001) Biomineralization, principles and concepts in bioinorganic materials chemistry. Oxford University Press
33. Bäuerlein E (2000) Biomineralization. Wiley, Weinheim
34. Mann S, Ozin GA (1996) Nature 382:313
35. Mann S (2000) Angew Chem Int Ed 39:3392
36. Dujardin E, Mann S (2002) Adv Mater 14:775
37. Ozin GA (1997) Acc Chem Res 30:17
38. Ozin GA (2000) Chem Commun, p 419
39. Kato T, Sugawara A, Hosoda N (2002) Adv Mater 14:869
40. Cölfen H, Mann S (2003) Angew Chem Int Ed 42:2350
41. Mann S (1996) Biomimetic materials chemistry. VCH, New York
42. Mann S (1993) Nature 365:499
43. Davis SA, Breulmann M, Rhodes KH, Zhang B, Mann S (2001) Chem Mater 13:3218
44. Cölfen H (2003) Curr Opin Colloid Interface Sci 8:23
45. Meldrum FC (2003) Int Mater Rev 48:187
46. Weiner S, Addadi L (1997) J Mater Chem 7:689
47. Mann S (1997) J Chem Soc, Dalton Trans, p 3953
48. van Bommel KJC, Friggeri A, Shinkai S (2003) Angew Chem Int Ed 42:980
49. Bäuerlein E (2003) Angew Chem Int Ed 42:614
50. Belcher AM, Wu XH, Christensen RJ, Hansma PK, Stucky GD, Morse DE (1996) Nature 381:56
51. Beniash E, Aizenberg J, Addadi L, Weiner S (1997) J Roy Soc Lond B 264:461
52. Beniash E, Addadi L, Weiner S (1999) J Struct Biol 125:50
53. Aizenberg J, Lambert G, Weiner S, Addadi L (2002) J Am Chem Soc 124:32

54. Addadi L, Raz S, Weiner S (2003) Adv Mater 15:959
55. Bäuerlein E (2004) Biomineralization, progress in biology, molecular biology and application, 2nd edn. Wiley, Weinheim
56. Politi Y, Arad T, Klein E, Weiner S, Addadi L (2004) Science 306:1161
57. Yu SH, Cölfen H (2004) J Mater Chem 14:2124
58. Wulff G (1901) Z Krystallogr 34:449
59. Rieger J, Hädicke E, Rau IU, Boeckh D (1997) Tens Surf Deterg 34:430
60. Söhnel O, Mullin JW (1982) J Cryst Growth 60:239
61. Yu SH, Cölfen H, Antonietti M (2003) J Phys Chem B 107:7396
62. Bolze J, Peng B, Dingenouts N, Nanine P, Narayanan T, Ballauff M (2002) Langmuir 18:8364
63. Bolze J, Pontoni D, Ballauff M, Narayanan T, Cölfen H (2004) J Colloid Interface Sci 277:84
64. Arnott HJ (1980) Mechanisms of biomineralization in animals and plants. Tokai University Press
65. Aizenberg J, Lambert G, Addadi L, Weiner S (1996) Adv Mater 8:222
66. Aizenberg J, Lambert G, Weiner S, Addadi L (2002) J Am Chem Soc 124:32
67. Taylor MG, Simkiss K, Greaves GN, Okazaki M, Mann S (1993) Proc Soc Lond B 252:75
68. Levi-Kalisman Y, Raz S, Weiner S, Addadi L (2002) Adv Funct Mater 12:43
69. Hasse B, Ehrenberg H, Marxen JC, Becker W, Epple M (2000) Chem Eur J 6:3679
70. Marxen JC, Becker W, Finke D, Hasse B, Epple M (2003) J Molluscan Stud 69:113
71. Weiss IM, Tuross N, Addadi L, Weiner S (2002) J Exp Zool 293:478
72. Nassif N, Pinna N, Gehrke N, Antonietti M, Jäger C, Cölfen H (2005) PNAS 102:12653
73. Raz S, Weiner S, Addadi L (2000) Adv Mater 12:38
74. Lowenstam HA, Weiner S (1985) Science 227:51
75. Ziegler A (1994) J Struct Biol 112:110
76. Tsortos A, Ohki S, Zieba A, Baier RE, Nancollas GH (1996) J Colloid Interf Sci 177:257
77. Mann S, Archibald DD, Didymus JM, Douglas T, Heywood BR, Meldrum FC, Reeves NJ (1993) Science 261:1286
78. Weiner S, Traub W (1984) Phil Trans R Soc Lond B 304:425
79. Koga N, Nakagoe YZ, Tanaka H (1998) Thermochim Acta 318:239
80. Loste E, Wilson RM, Seshadri R, Meldrum FC (2003) J Cryst Growth 254:206
81. Sawada K (1997) Pure Appl Chem 69:921
82. Abdel-Aal N, Sawada K (2003) J Cryst Growth 256:188
83. Brecevic L, Nielsen AE (1989) J Cryst Growth 98:504
84. Clarkson JR, Price TJ, Adams CJ (1992) J Chem Soc, Farady Trans 88:243
85. Falini G, Albeck S, Weiner S, Addadi S (1996) Science 271:67
86. Orme CA, Noy A, Wierzbicki A, McBride MT, Grantham M, Teng HH, Dove PM, DeYoreo JJ (2001) Nature 411:775
87. Burton WK, Cabrera N, Frank FC (1951) Phil Trans R Soc Lond A 243:299
88. Wegner G, Baum P, Müller M, Norwig J, Landfester K (2001) Macromol Symp 175:349
89. Munoz-Espi R, Qi Y, Lieberwirth I, Gomez CM, Wegner G (2005) Chem Eur J 12:118
90. Lu CH, Qi LM, Cong HL, Wang XY, Yang JH, Yang LL, Zhang DY, Ma JM, Cao WX (2005) Chem Mater 17:5218
91. Hetherington NBJ, Kulak AN, Sheard K, Meldrum FC (2006) Langmuir 22:1955
92. DeOliveira DB, Laursen RA (1997) J Am Chem Soc 119:10627

93. Sun Y, Mayers B, Herricks T, Xia Y (2003) Nano Lett 3:955
94. Penn RL, Banfield JF (1999) Geochim Cosmochim Acta 63:1549
95. Banfield JF, Welch SA, Zhang H, Ebert TT, Penn RL (2000) Science 289:751
96. Alivisatos AP (2000) Science 289:736
97. Penn RL, Banfield JF (1998) Science 281:969
98. Polleux J, Pinna N, Antonietti M, Niederberger M (2004) Adv Mater 16:436
99. Polleux J, Pinna N, Antonietti M, Hess C, Wild U, Schlögl R, Niederberger M (2005) Chem Eur J 11:3541
100. Penn RL, Oskam G, Strathmann TJ, Searson PC, Stone AT, Veblen DR (2001) J Phys Chem B 105:2177
101. Penn RL, Stone AT, Veblen DR (2001) J Phys Chem B 105:4690
102. Giersig M, Pastoriza-Santos I, Liz-Marzahn LM (2004) J Mater Chem 14:607
103. Pacholski C, Kornowski A, Weller H (2002) Angew Chem Int Ed 41:1188
104. Huang F, Zhang HZ, Banfield JF (2003) Nano Lett 3:373
105. Penn RL (2004) J Phys Chem B 108:12707
106. Gehrke N, Cölfen H, Pinna N, Antonietti M, Nassif N (2005) Cryst Growth Design 5:1317
107. Yang HG, Zeng HC (2004) Angew Chem Int Ed 43:5930
108. Antonietti M, Breulmann M, Göltner CG, Cölfen H, Wong KKW, Walsh D, Mann S (1998) Chem Eur J 4:2493
109. Qi LM, Cölfen H, Antonietti M, Li M, Hopwood JD, Ashley AJ, Mann S (2001) Chem Eur J 7:3526
110. Yu SH, Cölfen H, Antonietti M (2002) Chem Eur J 8:2937
111. Yu SH, Antonietti M, Cölfen H, Hartmann J (2003) Nano Lett 3:379
112. Cölfen H (2006) Non-classical crystallization. In: Arias JL, Fernandez MS (eds) Biomineralization: from paleontology to materials science. Editorial Universitaria, Universidad de Chile, Santiago, Chile (in press)
113. Cölfen H, Antonietti M (2005) Angew Chem Intl Ed 44:5576
114. Rousseau M, Lopez E, Stempfle P, Brendle M, Franke L, Guette A, Naslain R, Bourrat X (2005) Biomaterials 26:6254
115. Oaki Y, Imai H (2005) Angew Chem Int Ed 44:6571
116. Oaki Y, Imai H (2006) Small 2:66
117. Busch S, Dolhaine H, DuChesne A, Heinz S, Hochrein O, Laeri F, Podebrad O, Vietze U, Weiland T, Kniep R (1999) Eur J Inorg Chem 10:1643
118. Kniep R, Busch S (1996) Angew Chem Int Ed 35:2624
119. Busch S, Schwarz U, Kniep R (2003) Adv Funct Mater 13:189
120. Simon P, Carillo-Cabrera W, Formanek P, Göbel C, Geiger D, Ramlau R, Tlatlik H, Buder J, Kniep R (2004) J Mater Chem 14:2218
121. Simon P, Schwarz U, Kniep R (2005) J Mater Chem 15:4992
122. Wang TX, Cölfen H, Antonietti M (2005) J Amer Chem Soc 127:3246
123. Wohlrab S, Pinna N, Antonietti M, Cölfen H (2005) Chem Eur J 10:2903
124. Jongen N, Bowen P, Lemaitre J, Valmalette JC, Hofmann H (2000) J Colloid Interf Sci 226:189
125. Soare LC (2004) PhD thesis 3083, Ecole Polytechnique Federale de Lausanne (Lausanne)
126. Pujol O, Bowen P, Stadelmann PA, Hofmann H (2004) J Phys Chem B 108:13128
127. Nassif N, Gehrke N, Pinna N, Shirshova N, Tauer K, Antonietti M, Cölfen H (2005) Angew Chem Int Ed 44:6004
128. Weiner S, Addadi L, Wagner HD (2000) Mater Sci Eng C 11:1

129. Sethmann I, Hinrichs R, Wörheide G, Putnis A (2006) J Inorg Biochem 100:88
130. Sethmann I, Putnis A, Grassmann O, Löbmann P (2005) Am Mineral 90:1213
131. Weiner S, Sagi I, Addadi L (2005) Science 309:1027
132. Gehrke N, Nassif N, Pinna N, Antonietti M, Gupta HS, Cölfen H (2005) Chem Mater 17:6514
133. Kuznetsov YG, Malkin AJ, McPherson A (2001) J Cryst Growth 232:30
134. Brooks R, Clark LM, Thurston EF (1950) Philos Trans Roy Soc London Ser A 243:145
135. Hunt TS (1866) Amer J Sci Arts 42:49
136. Lengyel VE (1937) Z Kristallogr 97:67
137. Gower LA, Tirrell DA (1998) J Cryst Growth 191:153
138. Hardikar VV, Matijević E (2001) Colloids Surf A 186:23
139. Addadi L, Weiner S (1992) Angew Chem Int Ed 31:153
140. Gauldie RW, Nelson DGA (1988) Comp Biochem Physiol A 90:501
141. Wohlrab S, Cölfen H, Antonietti M (2005) Angew Chem Int Ed 44:4087
142. Faatz M, Gröhn F, Wegner G (2004) Adv Mater 16:996
143. Faatz M, Gröhn F, Wegner G (2005) Mat Sci Engin C 25:153
144. Patel VM, Sheth P, Kurz A, Ossenbeck M, Shah DO, Gower LB (2002) Synthesis of calcium carbonate-coated emulsion droplets for drug detoxification. In: Markovic B, Somansundaran P (eds) Concentrated colloidal dispersions: theory, experiments, and applications. ACS Symposium Series 878. American Chemical Society, Washington DC
145. Olszta MJ, Odom DJ, Douglas EP, Gower LB (2003) Conn Tissue Res 44(Suppl. 1):326
146. Antonietti M (2001) Curr Opin Colloid Interf Sci 6:244
147. Cölfen H (2001) Macromol Rapid Commun 22:219
148. Arias JL, Fernandez MS (2003) Mater Character 50:189
149. Sanchez C, Arribart H, Giraud Guille MM (2005) Nature Mater 4:277
150. Tang ZY, Kotov NA (2005) Adv Mater 17:951
151. Cölfen H, Yu SH (2005) MRS Bulletin 30:727
152. Naka K, Chujo Y (2001) Chem Mater 13:3245
153. Luquet G, Marin F (2004) CR Palevol 3:515
154. Kröger N, Deutzmann R, Sumper M (1999) Science 286:1129
155. Ritchie HH, Wang LH (1996) J Biol Chem 271:21695
156. Ritchie HH, Wang LH, Knudtson K (2001) Biochim Biophys Acta 1520:212
157. Gu KN, Chang SR, Slaven MS, Clarkson BH, Rutherford RB, Ritchie HH (1998) Eur J Oral Sci 106:1043
158. Sarashina I, Endo K (2001) Marine Biotechnol 3:362
159. Gotliv BA, Addadi L, Weiner S (2003) Chem Bio Chem 4:522
160. Gotliv BA, Kessler N, Sumerel JL, Morse DE, Tuross N, Addadi L, Weiner S (2005) Chem Bio Chem 6:304
161. Arias JL, Neira-Carrillo A, Arias JI, Escobar C, Bodero M, David M, Fernandez MS (2004) J Mater Chem 14:2154
162. Gerbaud V, Pignol D, Loret E, Bertrand JA, Berlandi Y, Fontecilla-Camps JC, Canselier JP, Gabas N, Verdier JM (2000) J Biol Chem 275:1057
163. Kilian CE, Wilt FH (1996) J Biol Chem 271:9150
164. Wilt FH (1999) J Struct Biol 126:216
165. MacDougall M, Simmons D, Luan X, Nydegger J, Feng J, Gu TT (1997) J Biol Chem 272:835
166. Mann K, Weiss IM, Andre S, Gabius HJ, Fritz M (2000) Eur J Biochem 267:5257
167. Evans JS (2003) Curr Opin Colloid Interface Sci 8:48

168. Jia Z, Davies PL (2002) Trends Biochem Sci 27:101
169. Yang DSC, Hon WC, Bubanko S, Xue Y, Seetharaman J, Hew CJ, Sicheri F (1998) Biophys J 74:2142
170. Jiaz Z, DeLuca CI, Chao H, Davies PL (1996) Nature 384:285
171. Marin F, Corstjens P, De Gaulejac B, De Vrind-De Jong E, Westbroek P (2000) J Biol Chem 275:20667
172. Arias JL, Carrino DA, Fernandez MS, Rodriguez JP, Dennis JE, Caplan AI (1992) Arch Biochem Biophys 298:293
173. Fernandez MS, Araya M, Arias JL (1997) Matrix Biol 16:13
174. Fernandez MS, Moya A, Lopez L, Arias JL (2001) Matrix Biol 20:793
175. Dennis JE, Carrino DA, Yamashita K, Caplan AI (2000) Matrix Biol 19:683
176. Aizenberg J, Black AJ, Whitesides GM (1999) Nature 398:495
177. Briseno AL, Aizenberg J, Han YJ, Penkala RA, Moon H, Lovinger AJ, Kloc C, Bao ZA (2005) J Amer Chem Soc 127:12164
178. Park RJ, Meldrum FC (2002) Adv Mater 14:1167
179. Park RJ, Meldrum FC (2004) J Mater Chem 14:2291
180. Shen FH, Feng QL, Wang CM (2002) J Crystal Growth 242:239
181. Sondi I, Salopek-Sondi B (2005) Langmuir 21:8876
182. Feng QL, Pu G, Pei Y, Cui FZ, Li HD, Kim TN (2000) J Crystal Growth 216:459
183. Raz S, Weiner S, Addadi L (2000) Adv Mater 12:38
184. Li CM, Botsaris GD, Kaplan DL (2002) Cryst Growth Des 2:387
185. Jiang HD, Liu XY, Zhang G, Li Y (2005) J Biol Chem 280:42061
186. Leveque I, Cusack M, Davis SA, Mann S (2004) Angew Chem Int Ed 43:885
187. Mirkin CA (2000) Inorg Chem 39:2258
188. Mirkin CA, Letsinger RL, Mucic RC, Storhoff JJ (1996) Nature 382:607
189. Kumar A, Pattarkine M, Bhadbhade M, Mandale AB, Ganesh KN, Datar SS, Dharmadhikari CV, Sastry M (2001) Adv Mater 13:341
190. Warner MG, Hutchison JE (2003) Nature Mater 2:272
191. Harnack O, Ford WE, Yasuda A, Wessels JM (2002) Nano Lett 2:919
192. Keren K, Krueger M, Gilad R, Ben Yoseph G, Sivan U, Braun E (2002) Science 297:72
193. Braun E, Eichen Y, Sivan U, Ben Yoseph G (1998) Nature 391:775
194. Richter J, Seidel R, Kirsch R, Mertig M, Pompe W, Plaschke J, Schackert HK (2000) Adv Mater 12:507
195. Ford WE, Harnack O, Yasuda A, Wessels JM (2001) Adv Mater 13:1793
196. Torimoto T, Yamashita M, Kuwabata S, Sakata T, Mori H, Yoneyama H (1999) J Phys Chem B 103:8799
197. Jin J, Jiang L, Chen X, Yang WS, Li TJ (2003) Chin J Chem 21:208
198. Alivisatos P (2004) Nature Biotechnol 22:47
199. Alivisatos AP, Johnson K, Peng X, Wilson TE, Loweth CJ, Bruchez M, Schultz PG, Nature 382:609
200. Dujardin E, Hsin LB, Wang CRC, Mann S (2001) Chem Commun 14:1264
201. Lee SW, Lee SK, Belcher AM (2003) Adv Mater 15:689
202. Li M, Mann S (2004) J Mater Chem 14:2260
203. Sadasivan S, Dujardin E, Li M, Johnson CJ, Mann S (2005) Small 1:103
204. Fu X, Wang Y, Huang L, Sha Y, Gui L, Lai L, Tang Y (2003) Adv Mater 15:902
205. Ryadnov MG, Woolfson DN (2004) J Amer Chem Soc 126:7454
206. Walsh D, Arcelli L, Ikoma T, Tanaka J, Mann S (2003) Nature Mater 2:386
207. Cao Y, Zhou YM, Shan Y, Huang XJ, Xue XJ, Wu ZH (2004) Adv Mater 16:1189
208. Mao C, Flynn CE, Hayhurst A, Sweeney R, Qi J, Georgiou G, Iverson B, Belcher AM (2003) Proc Natl Acad Sci 100:6946

209. Lee, Mao, Flynn, Belcher (2002) Science 296:892
210. Mao C, Solis DJ, Reiss BD, Kottmann ST, Sweeney RY, Hayhurst A, Georgiou G, Iverson B, Belcher AM (2004) Science 303:213
211. Patil AJ, Muthusamy E, Mann S (2004) Angew Chem Int Ed 43:4928
212. Patil AJ, Muthusamy E, Mann S (2005) J Mater Chem 15:3838–3843
213. Ma D, Li M, Patil AJ, Mann S (2004) Adv Mater 16:1838–1841
214. Rautaray D, Kumar A, Reddy S, Sainkar SR, Sastry M (2002) Cryst Growth Design 2:197
215. Yang JH, Qi LM, Zhang DB, Ma JM, Cheng HM (2004) Cryst Growth Design 4:1371
216. Walsh D, Kulak A, Aoki K, Ikoma T, Tanaka J, Mann S (2004) Angew Chem Int Ed. 43:6691
217. Walsh D, Aracelli L, Ikoma T, Tanaka J, Mann S (2003) Nature Mater 2:386
218. Gonzalez-McQuire R, Green D, Walsh D, Hall S, Chane-Ching JY, Oreffo ROC, Mann S (2005) Biomaterials 26:6652
219. Yu SH, Cui XJ, Li LL, Li K, Yu B, Antonietti M, Cölfen H (2004) Adv Mater 16:1636
220. Jones F, Cölfen H, Antonietti M (2000) Colloid Polym Sci 278:491
221. Jones F, Cölfen H, Antonietti M (2000) Biomacromolecules 1:556
222. Kröger N, Deutzmann R, Bergsdorf C, Sumper M (2000) PNAS 97:14133
223. Sumper M (2002) Science 295:2430
224. Kröger N, Lorenz S, Brunner E, Sumper M (2002) Science 298:584
225. Naik RR, Whitlock PW, Rodriguez F, Brott LL, Glawe DD, Clarson SJ (2003) Chem Commun 2:238
226. Menzel H, Horstmann S, Behrens P, Bärnreuther P, Krueger I, Jahns M (2003) Chem Commun 24:2994
227. Patwardhan SV, Mukherjee N, Steinitz-Kannan M, Clarson SJ (2003) Chem Commun 10:1122
228. Peytcheva A, Cölfen H, Antonietti M (2002) Colloid Polym Sci 280:218
229. Gower LB, Odom DJ (2000) J Crystal Growth 210:719
230. Yu SH, Cölfen H, Xu AW, Dong WF (2004) Cryst Growth Design 4:33
231. Bigi A, Boanini E, Walsh D, Mann S (2002) Angew Chem Int Ed 41:2163
232. Cheng XG, Gower LB (2006) Biotechnol Prog 22:141
233. Olszta MJ, Gajjeraman S, Kaufman M, Gower LB (2004) Chem Mater 16:2355
234. Olszta MJ, Douglas EP, Gower LB (2003) Calc Tissue Int 72:583
235. Volkmer D, Harms M, Gower LB, Ziegler A (2005) Angew Chem Int Ed 44:639
236. Jada J, Verraes A (2003) Colloids Surfaces A 219:7
237. Yu JG, Liu SW, Cheng B (2005) J Cryst Growth 275:572
238. Yu JG, Yu JC, Zhang L, Wang XC, Wu L (2004) Chem Commun, p 2414
239. Liu SW, Yu JG, Cheng B, Zhao L, Zhang Q (2005) J Cryst Growth 279:461
240. Yu Q, Ou HD, Song RQ, Xu AW (2006) J Cryst Growth 286:178
241. Xu G, Yao N, Aksay IA, Groves JT (1998) J Amer Chem Soc 120:11977
242. Boggavarapu S, Chang J, Calvert P (2000) Mater Sci Eng C11
243. Rieger J (2002) Tens Surf Deter 39:221
244. Deng SG, Cao JM, Feng J, Guo J, Fang BQ, Zheng MB, Tao J (2005) J Phys Chem B 109:11473
245. Qi LM, Li J, Ma JM (2002) Adv Mater 14:300
246. Balz M, Therese HA, Li J, Gutmann JS, Kappl M, Nasdala L, Hofmeister W, Butt HJ, Tremel W (2005) Adv Funct Mater 15:683
247. DiMasi E, Patel VM, Sivakumar M, Olszta MJ, Yang YP, Gower LB (2002) Langmuir 18:8902

248. Sugawara T, Suwa Y, Ohkawa K, Yamamoto H (2003) Macromol Rapid Commun 24:847
249. Oaki Y, Imai H (2004) Angew Chem 43:1363
250. Oaki Y, Imai H (2005) Langmuir 21:863
251. Oaki Y, Imai H (2004) J Amer Chem Soc 126:9271
252. Oaki Y, Imai H (2005) Adv Funct Mater 15:1407
253. Oaki Y, Imai H (2005) Chem Commun 48:6011
254. Aoki K, Mann S (2005) J Mater Chem 15:111
255. Sun Q, Beelen TPM, van Santen RA, Hazelaar S, Vrieling EG, Gieskes WWC (2002) J Phys Chem B 106:11539
256. Luo LB, Yu SH, Qian HS, Zhou T (2002) J Amer Chem Soc 127:2822
257. Gong JY, Luo LB, Yu SH, Qian HS, Fei LF (2006) J Mater Chem 16:101
258. Zhao M, Sun L, Crooks RM (1998) J Amer Chem Soc 120:4877
259. Zhao M, Crooks RM (1999) Adv Mater 11:217
260. Balogh L, Tomalia DA (1998) J Am Chem Soc 120:7355
261. Strable E, Bulte JWM, Moskowitz B, Vivekanandan K, Allen Mand T Douglas (2001) Chem Mater 13:2201
262. Garcia ME, Baker LA, Crooks (1999) Anal Chem 71:256
263. Sooklal K, Hanus LH, Ploehn HJ, Murphy CJ (1998) Adv Mater 10:1083
264. Gröhn F, Bauer BJ, Akpalu YA, Jackson CL, Amis EJ (2000) Macromolecules 33:6042
265. Keki S, Torok J, Deak G, Daroczi L, Zsuga M (2000) J Colloid Interface Sci 229:550
266. Esumi K, Hosoya T, Suzuki A, Torigoe K (2000) J Colloid Interface Sci 226:346
267. Gröhn F, Kim G, Bauer AJ, Amis EJ (2001) Macromolecules 34:2179
268. Hedden RC, Bauer BJ, Smith AP, Gröhn F, Amis EJ (2002) Polymer 43:5473
269. Esumi K, Suzuki A, Aihara N, Usui K, Torigoe K (1998) Langmuir 14:3157
270. Esumi K, Suzuki A, Yamahira A, Torigoe K (2000) Langmuir 16:2604
271. Naka K, Tanaka Y, Chujo Y, Ito Y (1999) Chem Commun 19:1931
272. Naka K (2003) Mater Life Sci Top Curr Chem 228:141
273. Naka K, Tanaka Y, Chujo Y (2002) Langmuir 18:3655
274. Donners JJJM, Heywood BR, Meijer EW, Nolte RJM, Roman C, Schenning APLHJ, Sommerdijk NAJM (2000) Chem Commun 19:1937
275. Donners JJJM, Heywood BR, Meijer EW, Nolte RJM, Sommerdijk NAJM (2002) Chem Eur J 8:2561
276. Förster S, Plantenberg T (2002) Angew Chem Int Ed 41:689
277. Kriesel JW, Sander MS, Tilley TD (2001) Chem Mater 13:3554
278. Napper DH (1989) Polymeric stabilization of colloidal dispersions. Academic, London
279. Tjandra W, Yao J, Ravi P, Tam KC, Alamsjah A (2005) Chem Mater 17:4865
280. Yu SH, Cölfen H, Mastai Y (2004) J Nanosci Nanotech 4:291
281. Guillemet B, Faatz M, Gröhn F, Wegner G, Gnanou Y (2006) Langmuir 22:1875
282. Cölfen H, Qi LM (2001) Chem Eur J 7:106
283. Antonietti M (2003) Nature Mater 2:9
284. Sedlak M, Antonietti M, Cölfen H (1998) Macromol Chem Phys 199:247
285. Sedlak M, Cölfen H (2001) Macromol Chem Phys 202:587
286. Cölfen H, Antonietti M (1998) Langmuir 14:582
287. Marentette JM, Norwig J, Stockelmann E, Meyer WH, Wegner G (1997) Adv Mater 9:647
288. Norwig J (1997) Mol Cryst Liq Cryst 313:115

289. Yu SH, Cölfen H, Hartmann J, Antonietti M (2002) Adv Funct Mater 12:541
290. Rudloff J, Antonietti M, Cölfen H, Pretula J, Kaluzynski K, Penczek S (2002) Macromol Chem Phys 203:627
291. Kaluzynski K, Pretula J, Lapienis G, Basko M, Bartczak Z, Dworak A, Penczek S (2001) J Polym Sci A: Polym Chem 39:955
292. Rudloff J, Cölfen H (2004) Langmuir 20:991
293. Qi LM, Cölfen H, Antonietti M (2000) Angew Chem Int Ed 39:604
294. Qi LM, Cölfen H, Antonietti M (2000) Chem Mater 12:2392
295. Cölfen H, Qi LM, Mastai Y, Börger L (2002) Cryst Growth Des 2:191
296. Robinson KL, Weaver JVM, Armes SP, Marti ED, Meldrum FC (2002) J Mater Chem 12:890
297. Yu SH, Cölfen H, Antonietti M (2003) Adv Mater 15:133
298. Bagwell RB, Sindel J, Sigmund W (1999) J Mater Res 14:1844
299. Zhang DB, Qi LM, Ma JM, Cheng HM (2002) Chem Mater 14:2450
300. Öner M, Norwig J, Meyer WH, Wegner G (1998) Chem Mater 10:460
301. Taubert A, Palms D, Weiss O, Piccini MT, Batchelder DN (2002) Chem Mater 14:2594
302. Taubert A, Palms D, Glasser G (2002) Langmuir 18:4488
303. Taubert A, Kübel C, Martin DC (2003) J Phys Chem B 107:2660
304. Yu SH, Antonietti M, Cölfen H, Giersig M (2002) Angew Chem Int Ed 41:2356
305. Mastai Y, Sedlák M, Cölfen H, Antonietti M (2002) Chem Eur J 8:2430
306. Qi LM (2001) J Mater Sci Lett 20:2153
307. Mastai Y, Rudloff J, Cölfen H, Antonietti M (2002) Chem Phys Chem 3:119
308. Orth J, Meyer WH, Bellmann C, Wegner G (1997) Acta Polymer 48:490
309. Palmqvist LM, Lange FF, Sigmund W, Sindel J (2000) J Am Ceram Soc 83:1585
310. Sigmund WM, Sindel J, Aldinger F (1999) Zeitschrift für Metallkunde 90:990
311. Sindel J, Bell NS, Sigmund WM (1999) J Am Ceram Soc 82:2953
312. Sindel J, Sigmund W, Baretzky B, Aldinger F (1996) Untersuchungen zu Wechselwirkungen zwischen Bindemittel und Pulverteilchen in wässrigen Bariumtitanatschlickern. In: Ziegler G, Cherdron H, Hermel W, Hirsch J, Kolaska H (eds) Werkstoff- und Verfahrenstechnik. Proc Werkstoffwoche 6:617
313. Sigmund WM, Sindel J, Aldinger F (1997) Progr Colloid Polym Sci 105:23
314. Sindel J, Sigmund WM, Aldinger F (1998) Proc. IEKC, Stuttgart 6
315. Hoogeveen NG, Stuart MAC, Fleer GJ, Frank W, Arnold M (1996) Macromol Chem Phys 197:2553
316. Hoogeveen NG, Stuart MAC, Fleer GJ (1996) Colloids Surf A 117:77
317. Solomentseva IM, Stuart MAC, Fleer GJ (2000) Colloid J 62:218
318. Bijsterbosch HD, Stuart MAC, Fleer GJ, van Caeter P, Goethals EJ (1998) Macromolecules 31:7436
319. Bijsterbosch HD, Stuart MAC, Fleer GJ (1999) J Colloid Interf Sci 210:37
320. de Laat AWM, Bijsterbosch HD, Stuart MAC, Fleer GJ, Struijk CW (2000) Colloids Surf A 166:79
321. Creutz S, Jerome R (1999) Langmuir 15:7145
322. Yu SH, Cölfen H, Fischer A (2004) Colloids Surf A: Physicochem Eng Aspects 243:49
323. de Laat AWM, Schoo HFM (1998) Colloid Polym Sci 276:176
324. de Laat AWM, Schoo HFM (1997) J Colloid Interf Sci 191:416
325. de Laat AWM, Schoo HFM (1998) J Colloid Interf Sci 200:228
326. Thünemann AF, Schütt D, Kaufner L, Pison U, Möhwald H (2006) Langmuir 22:2351
327. Creutz S, Jerome R, Kaptijn GMP, van der Werf AW, Akkerman JM (1998) J Coat Technol 70:41

328. Bronstein LM, Sidorov SN, Gourkova AY, Valetsky PM, Hartmann J, Breulmann M, Cölfen H, Antonietti M (1998) Inorg Chim Acta 280:348
329. Sidorov SN, Bronstein LM, Valetsky PM, Hartmann J, Cölfen H, Schnablegger H, Antonietti M (1999) J Colloid Interface Sci 212:197
330. Bronstein LM, Sidorov SN, Valetsky PM, Hartmann J, Cölfen H, Antonietti M (1999) Langmuir 15:6256
331. Zhang D, Qi LM, Ma JM, Cheng H (2001) Chem Mater 13:2753
332. Qi LM, Cölfen H, Antonietti M (2002) Nano Lett 1:61
333. Bouyer F, Gérardin C, Fajula F, Putaux JL, Chopin T (2003) Colloids Surf A: Physicochem Eng Aspects 217:179
334. Gerardin C, Sanson N, Bouyer F, Fajula F, Putaux JL, Joanicot M, Chopin T (2003) Angew Chem Int Ed 42:3681
335. Gerardin C, Buissette V, Gaudemet F, Anthony O, Sanson N, DiRenzo F, Fajula F (2002) Mater Res Soc Symp Proc 726:Q7.5
336. Sanson N, Putaux JL, Destarac M, Gerardin C, Fajula F (2005) Macromol Symp 226:279
337. Sanson N, Bouyer F, Gerardin C, In M (2004) Phys Chem Chem Phys 6:1463
338. Semagina NV, Bykov AV, Sulman EM, Matveeva VG, Sidorov SN, Dubrovina LV, Valetsky PM, Kiselyova OI, Khokhlov AR, Stein B, Bronstein LM (2004) J Mol Catal A Chem 208:273
339. Chen SF, Yu SH, Wang TX, Jiang J, Cölfen H, Hu B, Yu B (2005) Adv Mater 17:1461
340. Richter A, Petzold D, Hofmann H, Ulrich B (1996) Chem Technik 48:271
341. Kasparova P (2002) PhD thesis, Potsdam University
342. Rieger J, Hädicke E, Rau IU, Boeckh D (1997) Tens Surf Deterg 34:430
343. Li M, Cölfen H, Mann S (2004) J Mater Chem 14:2269
344. Dalas E, Klepetsanis P, Koutsoukos PG (1999) 15:8322
345. Kniep R, Busch S (1996) Angew Chem Int Ed 35:2624
346. Busch S, Dolhaine H, DuChesne A, Heinz S, Hochrein O, Laeri F, Podebrad O, Vietze U, Weiland T, Kniep R (1999) Eur J Inorg Chem 10:1643
347. Busch S, Schwarz U, Kniep R (2003) Adv Funct Mater 13:189
348. Cölfen H, Qi LM (2001) Progr Colloid Polym Sci 117:200
349. Küther J, Seshadri R, Tremel W (1998) Angew Chem Int Ed 37:3044
350. Küther J, Seshadri R, Nelles G, Assenmacher W, Butt HJ, Mader W, Tremel W (1999) Chem Mater 5:1317
351. Kasparova P, Antonietti M, Cölfen H (2004) Colloids Surf A Phys Engin Aspects 250:153
352. Evans JS (2003) Curr Opin Colloid Interface Sci 8:48
353. Yu SH, Cölfen H, Tauer K, Antonietti M (2005) Nature Mater 4:51
354. Walsh D, Lebeau B, Mann S (1999) Adv Mater 11:324
355. Wie H, Shen Q, Zhao Y, Wang D, Xu D (2004) J Cryst Growth 260:511
356. Wie H, Shen Q, Zhao Y, Wang D, Xu D (2004) J Cryst Growth 260:545
357. Shi HT, Qi LM, Ma JM, Cheng H (2003) J Am Chem Soc 125:3450
358. Shi HT, Qi LM, Ma JM, Wu NZ (2005) Adv Funct Mater 15:442
359. Zhang D, Qi LM, Ma LM, Cheng H (2002) Adv Mater 14:1499
360. Qi LM, Li J, Ma JM (2002) Chem J Chinese Univ 23:1595
361. Chen SF, Yu SH, Jiang J, Li FQ, Liu YK (2006) Chem Mater 18:115
362. Wegner G, Baum P, Müller M, Norwig J, Landfester K (2001) Macromol Symp 175:349
363. Basko M, Kubisa P (2002) Macromolecules 35:8948
364. Basko M, Kubisa P (2004) J Polym Sci A: Polym Chem 42:1189

Bio-inspired Crystal Growth by Synthetic Templates

Shu-Hong Yu

Division of Nanomaterials and Chemistry, Hefei National Laboratory for Physical Sciences at Microscale, School of Chemistry & Materials,
University of Science and Technology of China, Jinzhai Road 96, 230026 Hefei, China
shyu@ustc.edu.cn

1	Introduction	80
2	Basic Principles of Crystal Growth	81
2.1	Crystal Growth Habit and Crystal Shape	81
2.2	Crystal Growth Mechanisms in Solution	83
3	Synthetic Template Controlled Crystal Growth	84
3.1	Biopolymers	85
3.2	Synthetic Polymers	86
3.2.1	Polyelectrolytes	86
3.2.2	Graft Copolymers	90
3.2.3	Block Copolymers	90
3.2.4	Dendrimers	95
3.2.5	Foldamers	96
3.2.6	Supramolecular Functional Polymer	96
4	Synergistic Effects of Crystal Growth Modifiers	96
4.1	Combination of Mixed Polymer	96
4.2	Combination of Polymer with Low Mass Surfactant Molecules	97
4.3	Crystallization in a Mixture of Solvents	98
5	Artificial Interfaces and Matrices for Crystallization	99
5.1	Monolayers as Interfaces	99
5.2	Biopolymer Matrix	103
5.3	Synthetic Polymer Matrix	105
5.4	Crystallization on Foreign External Templates	107
5.5	Crystallization on Patterned Surfaces	109
6	Summary and Outlook	111
	References	112

Abstract Recent advances in bio-inspired strategies for the controlled growth of inorganic crystals using synthetic templates will be overviewed. There are a huge number of additives with different functionalities which can influence crystal growth; however, we only focus on the controlled growth and mineralization of inorganic minerals using synthetic templates as crystal growth modifiers, including biopolymers and synthetic polymers. New trends in the area of crystallization and morphogenesis of inorganic and inorganic–organic hybrid materials will be reviewed, including synergistic effects of crystal growth modifiers in water and in a mixed solvent, and crystallization on artificial interfaces or within matrices. Combination of a synthetic template with a normal surfactant or crystallization in a mixed solution system makes it possible to access various inorganic crystals with complex form and unique structural features. Several different morphogenesis mechanisms of crystal growth, such as selective adsorption, mesoscopic transformations, and higher order assembly, will be discussed. In addition, crystallization on artificial interfaces including monolayers, biopolymer and synthetic polymer matrices for controlled crystal growth, and emerging crystallization on foreign external templates and patterned surfaces for creation of patterned crystals will also be overviewed.

Keywords Crystal growth · Crystallization · Biopolymer · Synthetic polymer · Interface · Matrix · Self-assembly

1
Introduction

Biominerals are well-known composites of inorganic and organic materials in the form of fascinating shapes and high ordered structures, which exist in Nature as, for example, oyster shells, corals, ivory, sea urchin spines, cuttlefish bone, limpet teeth, magnetic crystals in bacteria, and human bones, created by living organisms [1, 2]. During the past few decades, it has been one of the hottest research subjects in materials chemistry and its cutting-edge fields to explore new bio-inspired strategies for self-assembling or surface-assembling molecules or colloids to generate materials with controlled morphologies, unique structural specialty, and complexity [3–10, 12, 13]. Especially, learning from Nature on how to create superstructures resembling naturally existing biominerals with their unusual shapes and complexity has attracted a lot of attention [14–20]. Furthermore, exploration of rational methods for the synthesis of a rich family of functional inorganic crystals or hybrid inorganic/organic materials with specific size, shape, orientation, organization, complex form, and hierarchy has also attracted a lot of attention owing to their importance and potential applications in industry [21–32].

In recent years, several excellent reviews on biomineralization/crystallization with different viewpoints and focuses have been published [10, 33, 34]. For example, biomineralization of unicellular organisms with complex mineral structures from the viewpoint of molecular biology [33], generating new organized materials through the biospecific interaction and coupling of biomolecules with inorganic nanosized building blocks [34], producing single-crystal mosaics, nanoparticle arrays, and emergent nanostructures with complex form and hierarchy by mesoscale self-assembly and cooperative transformation, and reorganization of hybrid inorganic–organic building blocks [10]. In addition, advances in polymer controlled crystallization and directed crystal growth, biomimetic mineralization, and synthesis of mesoscale order in hybrid inorganic–organic materials via nanoparticle self-assembly have been highlighted very recently [35–38].

In this review, the latest advances in the synthetic template controlled growth and crystallization of various inorganic minerals by bio-inspired approaches will be overviewed. The review is organized into five parts. First, general principles of crystal growth will be summarized (Sect. 2). Second, an overview of the recent advances in bio-inspired crystal growth of different inorganic crystals under the control of diverse biopolymers and synthetic templates will be summarized (Sect. 3). Emerging crystallization approaches and synergistic effects of crystal growth modifiers such as by a combination of polymers, or the combination of a polymer with low-mass surfactant molecules, as well as crystallization in a mixture of solvents, will be overviewed (Sect. 4). The diverse crystallization events occurring on the interfaces/substrates or within artificial matrices, and on emerging patterned crystallization will be overviewed (Sect. 5). Finally, we will give a summary and perspectives on this active research area (Sect. 6).

2
Basic Principles of Crystal Growth

2.1
Crystal Growth Habit and Crystal Shape

It is well known that diverse crystal shapes of the same compound are due to the differences of the crystal faces in surface energy and external growth environment [39]. Basically, crystal shape is determined by crystal habit and branching growth [39–45]. Crystal habit is determined by the relative order of surface energies of different crystallographic faces of a crystal [39–41], and in fact the shape of a crystal is usually the outside embodiment of its intrinsic cell replication and amplification. Branching growth created by a diffusion effect [42, 43] also plays an alternative role in the crystal shape due to the fact that the consumption of the ions or molecules near the surface of a growing crystal will form a concentric diffusion field around the crystal [44, 45]. This makes the apexes of a polyhedral crystal, which protrude further into the region of higher concentration, grow faster than the central parts of facets, thus forming branches [46].

As early as 1901, Wulff described a thermodynamic treatment of the crystal shape changes based on an energy minimized total surface area [39]. It is nowadays well known that this purely thermodynamic treatment cannot always predict the crystal shape, because crystallization and crystal shape often also rely on kinetic effects and defect structures like screw dislocations or kinks etc. The specific adsorption of ions or organic additives to particular faces can inhibit the growth of these faces. Generally, the growth rate of a crystal face is usually related to its surface energy if the same growth mechanism acts on each face. The fastest crystal growth will occur in the direction perpendicular to the face with the highest surface energy in order to eliminate or reduce higher energy surfaces, while lower energy surfaces will become more exposed in area. Thus, the fast growing faces usually have high surface energies and finally they will vanish in the final shape (Fig. 1a).

Crystal growth habit can be modified when the relative order of surface energies can be changed or when crystal growth along certain crystallographic directions is selectively hindered by a crystal growth modifier [4]. In the pres-

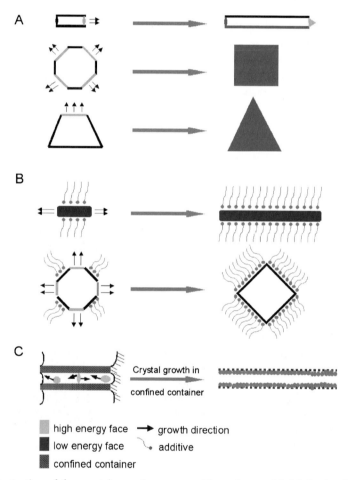

Fig. 1 Illustration of the crystal growth process. **a** Normal crystal habit in the absence of crystal modifiers; **b** the altered crystal growth habit due to the selective/preferential adsorption of crystal modifiers to a specific crystallographic face; and **c** the crystal growth within an confined environment

ence of crystal growth modifiers, the preferential/selective adsorption of crystal modifiers to a specific crystallographic face becomes stronger than that of others due to the anisotropy in adsorption stability decreasing the surface energy of the adsorbed face and inhibiting the crystal growth perpendicular to this face, thus altering the final shape of the crystal (Fig. 1b) [32, 47]. In addition, the crystal shape can be altered if the growth process occurs in a confined environment (Fig. 1c). The general crystal growth mechanisms in solution will be discussed in the following section.

2.2
Crystal Growth Mechanisms in Solution

The crystal growth mechanisms in solution are rather complicated and there are several that dominate the crystal growth process in the solution system. Recently, the classical and nonclassical crystallization mechanisms of inorganic minerals in solution systems have been described according to the latest developments in the crystallization field, as illustrated in Fig. 2 [35].

Primary nanoparticles are nucleated from the clusters called "critical crystal nuclei" which are the smallest crystalline units capable of further growth. Further growth of these primary nanoparticles by ion attachment and unit cell replication results in the formation of a final macroscopic single crystal, which is more or less an amplification of the initial crystal (Fig. 2a) [35].

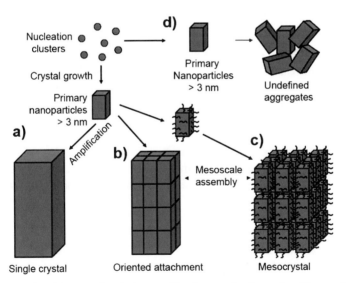

Fig. 2 **a** Classical and **b, c** nonclassical crystallization mechanisms via self-assembly. Crystallization starts from nucleation clusters (*upper left*). **d** The primary nanoparticles can also grow by ion attachment, but they then aggregate uncontrollably at a certain stage, forming undefined polycrystalline aggregates. (Reproduced from [35], © 2005, MRS)

These primary nanoparticles can also grow further into a single crystal by so-called oriented attachment of these nanoparticles [48–52] because their surfaces contain face-specific information (Fig. 2b). In addition, face-specific interactions between these primary building units can undergo directed self-assembly, resulting in a so-called mesocrystal [53], which is usually an intermediate on the formation pathway of a single crystal (Fig. 2c). These primary units can also grow by ion attachment alone, with unit cell replication, but they then aggregate at a certain stage, forming undefined polycrystalline aggregates (Fig. 2d).

Usually, more complex and emergent superstructures cannot be simply grown/constructed via a simple unit cell amplification process; instead they are normally formed via spontaneous self-organization of nanobuilding units carried with face-specific information in a controlled way and complicated mesoscale transformation mechanisms. The shape of crystals can be altered by various additives, i.e., inorganic cations and anions, organic additives, or even solvents, which has been overviewed recently [54]. In the following sections, we only focus on the latest developments in the emerging field of bio-inspired crystal growth of various inorganic crystals under the control of synthetic templates, including biopolymers, diverse synthetic polymers and their synergistic effects, and crystallization on artificial interfaces or within artificial matrices.

3
Synthetic Template Controlled Crystal Growth

Bio-inspired approaches for mimicking the biomineralization process of biominerals have been intensively studied. Usually, the synthetic templates used for these approaches include biopolymers and synthetic polymers. There are a huge number of synthetic polymers used for this purpose, including polyelectrolytes, graft copolymers, block copolymers, dendrimers, and foldamers.

The principle of controlled crystallization and morphosynthesis using a soluble polymer as crystal growth modifier in a solution which does not form an assembly is completely different from the templating effects, such as artificial interfaces and matrices or patterned substrates, which will be discussed in Sect. 5. The structure setup and evolution certainly does not rely on transcription or a straightforward template effect, but relies on a synergistic effect of the mutual interactions between functional groups of the polymers and inorganic species, and the subsequent reconstruction by a self-assembly process [4, 10, 55].

In the following sections, we will discuss recent advances in the area of polymer controlled crystal growth and morphogenesis of various inorganic minerals by the use of biopolymers, various synthetic polymers, and non-classical block copolymers—-so-called double-hydrophilic block copolymers (DHBCs) [35–37].

3.1
Biopolymers

Biopolymers consist of unique self-assembled structures and thus are often used as a natural soluble additive for the morphogenesis of complex superstructures. So far, in vitro experiments have failed to reproduce the complex biomineral shapes such as $CaCO_3$ by the use of a variety of additives such as dextran [56], collagen [57], soluble mollusk shell proteins extracted from nacre [58, 59], soluble macromolecules extracted from coralline algae [60], soluble macromolecules extracted from the respective layers of a mollusk shell [61, 62], protein secondary structures [63], and peptides for $CaCO_3$ crystallization [64]. The reason is that the biomineralization process is very complicated and an insoluble matrix can also influence the crystallization location as a compartment (Sect. 5).

The concept of using biopolymers for crystallization/mineralization has been widely expanded to synthesize and assemble different kinds of other functional inorganic nanostructures in recent years, which has been overviewed recently [35, 38, 54, 65] with specific examples. Directed assembly and organization through the interaction of biomolecule templates, such as DNA [66–69], virus [70], tobacco mosaic virus [71], and peptide [72], with inorgainc species like Au and ZnS make it possible to access organized nanostructures with different shape and dimensionalities. Magnetic crystals such as tabular single-domain magnetite [73] and a regulated cubic shape [74] can be grown at room temperature in water with close to neutral pH by biological methods. Some kinds of magnetotactic bacteria can synthesize and align ferromagnetic mineral greigite, as demonstrated by Mann et al. [75]. Sastry and coworkers have used fungi and actinomycetes, which can produce CO_2 during their growth, to form metal carbonate [76–79]. The same group has used the extracts of the lemongrass plant and of *Azadirachta indica* leaf to react with aqueous noble metal ions to produce thin, flat, single-crystalline gold nanotriangles [80], and silver and bimetallic Au/Ag nanoparticles [81].

The complex morphology of the nanopatterned silica diatom cell walls has been found to be related to species-specific sets of polycationic peptides, so-called silaffins, which were isolated from diatom cell walls [82]. The morphologies of precipitated silica can be controlled by changing the chain lengths of the polyamines as well as by a synergistic action of long-chain polyamines and silaffins [83, 84]. It has been proposed that the delicate pattern formation in diatom shells can be explained by phase separation of silica solutions in the presence of these polyamines [85]. Various linear synthetic analogs of the natural active polyamines in biosilica formation can accelerate the silicic acid condensation even more than the above mentioned silaffins [86]. Block copolypeptide poly(L-cysteine$_{30}$-b-L-lysine$_{200}$) has been used for biosilification [87]. Similar to the silaffins, silicateins can also catalyze the formation of silica at ambient conditions, but from tetraethoxysilane (TEOS) rather than silicic acid [88].

3.2
Synthetic Polymers

3.2.1
Polyelectrolytes

Simple low molecular mass polyelectrolytes can electrosterically stabilize inorganic colloids. Besides this function, low molecular mass polyelectrolytes have been widely used as additives in the controlled growth of diverse inorganic materials [35, 38]. The addition of polyelectrolytes with strong inhibition ability can stabilize the amorphous nanobuilding blocks in the early stage, and then stimulate a mesoscale transformation [10] or act as a material depot in a dissolution–recrystallization process. Time-resolved study of the scale inhibition efficiency of polycarboxylates has shown that amorphous precursor particles were formed in the initial stages [89] and were also observed in other cases.

For example, poly(acrylic acid) or poly(aspartic acid) crystal growth modifiers or structure directing agents can be used to induce the formation of various kinds of complex and hierarchical superstructures, such as structured calcium phosphate [90], helical $CaCO_3$ [91, 92], complex spherical $BaCO_3$ superstructures [93], hollow octacalcium phosphate ($Ca_8H_2(PO_4)_6 \cdot H_2O$) [94], $BaSO_4$ [95] and $BaCrO_4$ fiber bundles, or superstructures with complex repetitive patterns [96] as shown in Fig. 3.

The formation mechanism of the complex structures under control of a polyelectrolyte has not yet been well understood. Unusual complex structures of calcite helices or hollow helices can be obtained in the presence of a chiral polyaspartate, which can produce helical protrusions or occasionally hollow helices as shown in Fig. 3a [91]. Gower and colleagues have proposed a so-called polymer-induced liquid precursor (PILP) process to illustrate the possible mechanism for the formation of the complex morphologies [92]. In this process, it is believed that the addition of small amounts of polymer (μg mL^{-1} range) to the crystallizing solution results in the formation of a liquid–liquid phase separation to precursor droplets, which can adapt complex shapes before they crystallize. Yet, the formation mechanism of such amazing spherulitic vaterite aggregates with helical structures and with spiral pits is still not clear. $CaCO_3$ mineralization in the presence of collagen via a PILP process can result in the formation of fibrous aggregates [97]. Recently, Gower et al. demonstrated that rhombohedral calcite crystals grown in the absence of polymeric additives can be introduced into the above PILP process

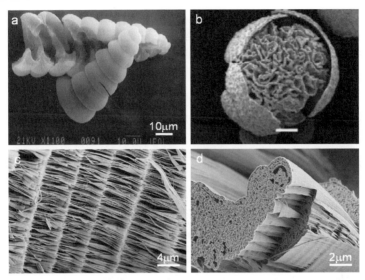

Fig. 3 Complex form of inorganic minerals formed by crystallization under control of a simple polyelectrolyte. **a** Helical CaCO$_3$ structures produced by the PILP process. Occasionally, the helices are partially hollow. The hollow helix fractured by micromanipulation. The *scale bar* represents 10 μm. (Reproduced from [91], © 1998, Elsevier Sciences). **b** Scanning electron microscopy (SEM) image of octacalcium phosphate (OCP)–polyelectrolyte architectures synthesized in the presence of polyaspartate, and isolated after different periods of aging, in solution for 3 h. The *scale bar* represents 100 μm. (Reproduced from [94], © 2002, Wiley). **c,d** Complex forms of BaSO$_4$ bundles and superstructures produced in the presence of 0.11 mM sodium polyacrylate (M_n = 5100), at room temperature, [BaSO$_4$] = 2 mM, pH = 5.3, 4 days. **c** The detailed superstructures with repetitive patterns. (Reproduced from [96], © 2003, American Chemical Society). **d** A zoomed SEM image of the well-aligned bundles. (Reproduced from [37], © 2005, MRS)

to act as seeds, and to induce the growth of crystalline calcite fibers on the surface by selective deposition of the PILP film on top of the rhombohedra rather than on the surrounding glass substrate [98] (Fig. 4). A liquid-phase mineral precursor, the so-called bobble head formed under physiological conditions (and down to temperatures as low as 4 °C), was observed on the tips of fibers, suggesting that an analogous solution–precursor–solid (SPS) mechanism may act in this system, which is quite similar to the vapor–liquid–solid (VLS) [99] and solution–liquid–solid (SLS) mechanisms [100] proposed for the growth of one-dimensional fibers under hot conditions.

The observed thin porous membrane of oriented octacalcium phosphate (OCP, Ca$_8$H$_2$(PO$_4$)$_6$·H$_2$O) crystals (Fig. 3b) on the outside of hollow OCP crystals mineralized in the presence of polyaspartate could be similar to that observed in the case of mineralization of CaCO$_3$ by a typical PILP process [92, 97].

Fig. 4 SEM micrographs of calcium carbonate deposited onto rhombohedral substrate crystals in the presence and absence of micromolar amounts of acidic polymer. **a** Calcite overgrowth on calcite substrates in the absence of polymeric additives. The *scale bar* represents 10 μm. **b** Calcite fibers grown on a solution-grown calcite substrate. In this case, the fibers appear to exhibit an isoepitaxial relationship with the underlying substrate crystal. The *scale bar* represents 20 μm. (Reproduced from [98], © 2004, American Chemical Society)

By crystallization of $BaSO_4$ and $BaCrO_4$ minerals using the sodium salt of poly(acrylic acid) as crystal modifier, elegant nanofiber bundles and their superstructures with conelike crystals [95] and hierarchical and repetitive growth patterns can be generated [96] (Fig. 3c,d), possibly based on a self-limiting growth mechanism. In this mechanism, a dipole crystal may be favored for a heterogeneous nucleation as one end of the crystal is determined by the heterogeneous surface instead of homogeneous solution and the other by the solution/dispersion [101]. A new heterogeneous nucleation event will occur on the rim of the mother crystal to start a self-limiting growth process, which is favorable for the formation of repeated patterns and a "cone-in-a-cone" superstructure (Fig. 3c,d). The low molecular weight polyelectrolytes poly(allylamine hydrochloride) (PAH) and poly(sodium 4-styrenesulfonate) (PSS) were also used for the mineralization of complex spherical $BaCO_3$ superstructures made of rodlike crystals [93], spherical $CaCO_3$ particles [102], and a hydroxyapatite/PAH–PSS polyelectrolyte composite shell [103]. Recently, unusual $CaCO_3$ superstructures, which transformed from the typical calcite rhombohedra to rounded edges, to truncated triangles, and finally to concavely bent lenslike superstructures using PSS as crystal modifier, were generated by a nonclassical crystallization process [104], where PSS can bind selectively to the otherwise nonexposed (001) calcite face, resulting in mesostructures composed of truncated triangular units instead of the typical rhombohedra.

Poly(L-isocyanoalanyl-D-alanine) with a regular distribution of carboxylic acid-terminated side chains was taken as a model template to investigate the relation between the structure of a polymeric template and a developing calcite phase [105]. A series of carboxylate-containing polyamides were synthesized [106–108] for the purpose of crystallization of $CaCO_3$. Helical calcite superstructures, with the helix turn corresponding to the copolymer enantiomer of chiral copolymers of phosphorylated serine (Ser) and aspartic acid (Asp) with molar masses 15 000–20 000 g mol^{-1}, could be generated under a limited experimental window [109] when a high degree of phosphorylated Ser (75 mol %) and 25 mol % Asp in the copolymer were applied.

Recently, CaCO$_3$ microspheres composed of vaterite nanoparticles with a size of 15–25 nm were mineralized by a simple polypeptide-directed strategy using sodium poly(aspartic acid) (M_w = 14 900) as an additive, and a remarkably soft nature of the nanoparticle assembly was found [110]. In addition, a family of superstructured vaterite mesocrystals, with hexagonal symmetry and uniform size and shape, could be mineralized by a vapor diffusion technique in the presence of an N-trimethylammonium derivative of hydroxyethyl cellulose [111]. Stable amorphous CaCO$_3$ hollow spheres have been synthesized using phytic acid as an additive [112], which is rich in phosphate groups and shows strong inhibiting ability for amorphous CaCO$_3$. The diameter of the amorphous CaCO$_3$ spheres could be controlled by adjusting the concentration of phytic acid [112].

Polyacrylamide (PAM) and carboxyl-functionalized polyacrylamide (PAM-COOH) were used as additives to selectively grow hexagonal ringlike ZnO structures and very thin discoid-like microstructures under mild conditions [113] (Fig. 5). The preferential and selective adsorption of polyacrylamide molecules on the (002) basal plane strongly inhibits the growth along this direction. In addition, the polymer occluded within the hybrid aggregates formed in the initial stage will undergo adsorption and desorption, and a polymer concentration gradient will form from the inside to the outside of the hybrid particle containing multiple chelating units. In this process, the dissolution of the core part of the crystal will occur to form ringlike struc-

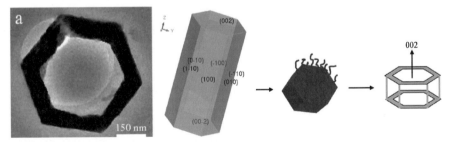

Fig. 5 a Transmission electron microscopy (TEM) image of a single ZnO nanoring formed in the presence of PAM. Modeling the morphology and growth habit of ZnO crystals in the absence of any external influence (*left*, simulated with the Cerius2 software), polymer adsorbed on (002) faces of ZnO (*middle*), and the final ZnO nanoring formation in the presence of PAM macromolecules (*right*). (Reproduced from [113], © 2006, American Chemical Society)

tures, as recently proposed in a block copolymer controlled crystallization of hollow CaCO$_3$ microrings (described in Sect. 3.2.3).

Imai et al. reported that poly(acrylic acid) (PAA) molecules with carboxy groups behave as a suppressant and template for the crystallization of calcium carbonate [114]. The results demonstrated that high molecular weight PAAs anchored to a glass surface promoted the oriented nucleation of calcite crystals. The combination of low and high molecular weight PAAs can achieve a moderate suppression effect, resulting in the formation of lozenge-shaped films consisting of iso-oriented crystal grains.

3.2.2
Graft Copolymers

Only a few examples have been demonstrated for crystallization control in aqueous environments using double hydrophilic graft copolymers [115] as reviewed previously [35]. Wegner and colleagues prepared poly(ethylene oxide) graft copolymers with methacrylic acid and/or vinylsulfonic acid by free radical polymerization [115], which have been applied to controlling the crystallization of ZnO [115]. In addition, poly(ethylene oxide) graft copolymers with a polyacetal backbone and functional carboxy groups along the main chain have been prepared [116], which were used as additives in the crystallization of CaCO$_3$ [117].

3.2.3
Block Copolymers

Block copolymers with hydrophilic and hydrophobic blocks show a similar behavior to low molecular weight surfactants [118, 119]. In recent years, various kinds of block copolymers have been used for controlled crystallization and stabilization of nanoparticles [35]. Among them, a new class of functional polymers, the so-called double-hydrophilic block copolymers (DHBCs), have been designed as crystal modifiers for mimicking the biomineralization process [35]. Typically, a DHBC consists of one hydrophilic block designed to interact strongly with the appropriate inorganic minerals and surfaces, and another hydrophilic block that does not interact (or only weakly interacts) and mainly promotes solubilization in water. Recently, progress has demonstrated that the DHBCs are very effective in crystallization control of various minerals [35, 36]. These polymers are designed to have typically rather small block lengths of $10^3 - 10^4$ g mol^{-1}. The solvating block shows good solubility in water and is in most cases a poly(ethylene oxide) (PEO) block and the binding block contains variable chemical patterns, which show strong affinity to minerals and have a strong interaction with inorganic crystals. The functional groups on the binding block usually include $-$OH, $-$COOH, $-$SO$_3$H, $-$SO$_4$, $-$PO$_3$H$_2$, $-$PO$_4$H$_2$, $-$SCN, $-$NR$_3$, $-$HNR$_2$, and $-$H$_2$NR [35].

Usually, DHBCs are used as excellent stabilizers for the in situ formation of various metal nanocolloids and semiconductor nanocrystals such as Pd, Pt [120–122], Au [120–123], Ag [124], CdS [125], and lanthanum hydroxide [126]. It has been shown that DHBCs [127, 128] can exert a strong influence on the morphogenesis of a variey of inorganic particles such as $CaCO_3$ [129–137], $BaCO_3$, $PbCO_3$, $CdCO_3$, $MnCO_3$ [136], calcium phosphate [138], barium sulfate [139–142], barium chromate [101, 143], barium titanate [144], calcium oxalate dihydrate [145], zinc oxide [132, 146–149], cadmium tungstate [150], and chiral organic crystals [151], and can act as a template for silica formation [152] and even control the structure of ice and water [153]. DHBCs can be used for the stabilization of specific planes of some crystals for their oriented growth such as Au [123], ZnO [146–148], calcium oxalate [145], $PbCO_3$ [136], and $BaSO_4$ [141].

Recent advances have demonstrated that DHBCs can act as an excellent stabilizer and crystal growth modifier for the formation of various inorganic crystals with interesting shapes, stabilize specific crystal faces, control crystal polymorphism by polymer adsorption, and mediate the mesoscopic transformation and higher order assembly of nanoparticles as reviewed recently [36–38]. In the following part, we emphasize the latest progress and new examples on the controlled growth of unusual superstructures of complex form, and illustrate the extraordinary ability, flexibility, and versatility of the DHBCs in the controlled growth of mineral superstructures.

A rigid DHBC-poly(ethylene glycol)-b-poly(1,4,7,10,13,16-hexaazacyclooctadecane ethylene imine) has been used as a crystal modifier to mineralize $CaCO_3$ crystals called "pancakes" with a shape similar to the layered struc-

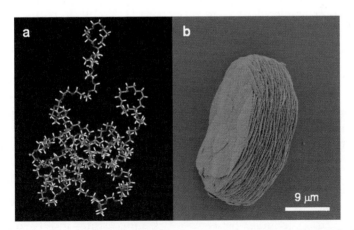

Fig. 6 a The structure of PEG-b-hexacyclen. **b** A typical SEM image of a pancake-like self-stacked $CaCO_3$ obtained after 2 weeks gas diffusion reaction in the presence of $1 g L^{-1}$ PEG-b-hexacyclen, starting pH 4, $[Ca^{2+}] = 10 mM$. (Reproduced from [154], © 2005, Wiley)

ture of nacre in *Haliotis rufescens*, as shown in Fig. 6 [154]. Layered crystals with different morphologies and surface structures, even disklike crystals, can be obtained by altering the mineralization conditions in the system. This study suggested that the morphologies of the crystal were not influenced by epitaxial match between the polymer and crystal faces, but that particle stabilization, crystallization time, time for polymer rearrangement, and surface ion density were of great effect on the resulting morphology [154].

Recently, a hydrophobically modified DHBC, poly(ethylene glycol)-*block*-poly(ethylene imine)-poly(acetic acid) (PEG-*b*-PEIPA), with on average one hydrophobic moiety at the end of the branched poly(ethylene imine) domain (PEG-*b*-PEIPA-C_{17}; PEG = $5000\,\mathrm{g\,mol^{-1}}$, PEIPA = $1800\,\mathrm{g\,mol^{-1}}$) [155] was found to form aggregates in aqueous solution, and showed an amazing texture and morphology control on the formation of unusual $CaCO_3$ microrings [156] (Fig. 7). It was proposed that the formation of $CaCO_3$ microrings is driven by the aggregation of polymer–inorganic hybrid nanocrystallites following crystallization inside unstructured polymer aggregates and the subsequent dissolution of nanocrystals from the inner side of the aggregates toward the outside. The hybrid core part of the disk is composed of primary nanocrystals with a large amount of attached polymer, which will have a great tendency to dissolve the $CaCO_3$ nanocrystals due to the large number of included multiple chelating ethylenediamine tetraacetic acid (EDTA) moieties from the block copolymer aggregates. This preferred dissolution event from the center of hybrid particles is similar to the formation of hollow $CaCO_3$ spheres under the control of a strongly chelating PEO-*b*-PEIPA additive, just without the hydrophobic tails [129]. This was also confirmed by a default experiment which showed that the rhombohedral calcite crystals formed in the absence of polymer were indeed dissolved in this polymer solution [156]. The formation of a polymer concentration gradient from the inside to the outside of the hybrid particle contained multiple chelating units, and the following restructuring within the hybrid structures, could be a reasonable and new explanation for such selective dissolution of the center of

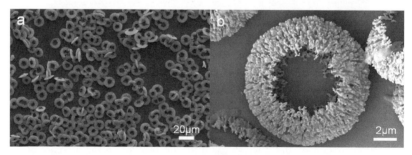

Fig. 7 SEM images of $CaCO_3$ microrings; $2.0\,\mathrm{g\,L^{-1}}$ PEG-*b*-PEIPA-C_{17}, 15 days, starting pH 4, [$CaCl_2$] = 20 mM. (Reproduced from [156], © 2006, American Chemical Society)

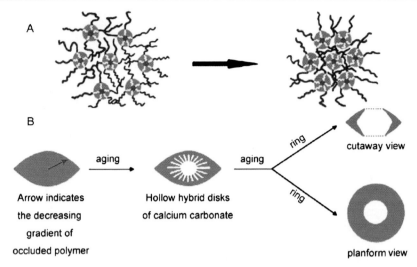

Scheme 1 a Aggregation mode of interlayer micelle-occluded nanocrystals. **b** Graphic presentation of the formation mechanism of CaCO$_3$ mesorings. *Gray*, calcium carbonate; *red*, hydrophobic block; *blue*, soluble neutral block; *yellow*, charged block. The *red arrow* indicates the decreasing gradient of occluded polymer within the disklike structure. (Reproduced from [156], © 2006, American Chemical Society)

the particles. This concept could also be helpful for the explanation of CaCO$_3$ hollow spheres [129] and amorphous CaCO$_3$ [112], and ZnO rings [113] by a polymer controlled crystallization process.

Recently, a racemic phosphonated DHBC, poly(ethylene glycol)-*b*-[2-(4-dihydroxyphosphoryl)-2-oxabutyl] acrylate ethyl ester (PEG-*b*-DHPOBAEE), was synthesized and applied to the controlled crystallization of BaCO$_3$ mineral. The chemical structure and a computer modeling structure of the functional head oligo[2-(4-dihydroxyphosphoryl)-2-oxabutyl] acrylate ethyl ester were depicted, showing its conformation of minimal energy in the absence of solvent (upper part in Fig. 8). It shows that this functional block is sterically overcrowded and adopts a stretched conformation, but with a high density of binding sites with mutual distances of 5–11 Å, implying a multiplicity of potential binding interactions with the inorganic crystal faces. Amazing BaCO$_3$ mineral helices have been successfully produced by programmed self-assembly of the elongated orthorhombic BaCO$_3$ units, based on the selective adsorption of PEG-*b*-DHPOBAEE onto the (110) faces of orthorhombic BaCO$_3$ (Fig. 8) [157]. This tectonic arrangement via coded self-assembly relies on two processes [157]: (1) the adsorption of the stiff DHBC onto the favorable (110) sites results in a staggered arrangement of aggregating nanoparticles that is controlled in direction after the aggregation of the first three particles, and (2) a particle approaching an aggregate in the perpendicular direction is presented with favorable and unfavorable adsorption

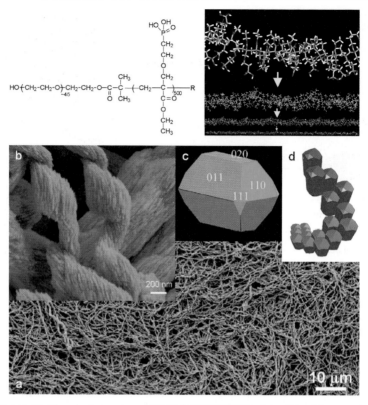

Fig. 8 *Top (left)*: Chemical structure of PEG-*b*-DHPOBAEE where R is either another PEG block or a hydrogen atom depending on the termination mode during polymerization. *Top (right)*: Different magnifications of the computer modeling results of the vacuum energy minimum conformation of the functional block of PEG-*b*-DHPOBAEE with 640 monomer units in vacuum. Note the stiff structure as a result of steric constraints. The modeling for the 640-mer was done with the *Cerius2* software (Accelrys). *Red*: O; *yellow*: P; *gray*: C; *white*: H. *Bottom*: **a** Helical BaCO$_3$ nanoparticle superstructures grown via a programmed self-assembly of elongated nanoparticles at room temperature using PEG-*b*-DHPOBAEE as template; 1 g L^{-1}, starting pH 4, [BaCl$_2$] = 10 mM. **b** Magnified SEM image showing the helical structure. **c** The primary nanocrystalline witherite building block in vacuum not representing observed face areas in solution but just illustrating the orientation of the relevant faces. **d** Proposed formation mechanism of the helical superstructure. (Reproduced from [157], © 2005, Nature Publishing Group)

sites, leading to a twist in the particle aggregate. The overlay of these two processes leads to helix formation (bottom part in Fig. 8). This successful access to inorganic helices demonstrated the possibility of selectively adsorbing additives onto specific crystal surfaces, and initiating the occurrence of the most advanced ways of so-called programmed self-assembly of nanoparticles, which can be used for the synthesis of various inorganic structures with unique structural features.

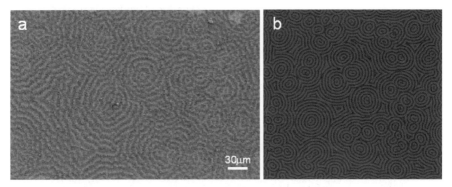

Fig. 9 a SEM images of the obtained concentric circle pattern of $BaCO_3$ crystals grown for 1 day; [polymer] = $1\,g\,L^{-1}$, $[Ba^{2+}]$ = 10 mM, starting pH = 5.5. **b** Simulation result with a modified Brüsselator model for the reaction–diffusion equations. (Reproduced from [158], © 2006, Wiley)

Very recently, the polymer PEG-*b*-DHPOBAEE has also been used for mineralization of $BaCO_3$ mineral to spontaneously form a concentric circle Belousov–Zhabotinsky pattern made of $BaCO_3$ nanorods in solution on a glass substrate [158]. The experimental evidence indicated that the formation of the Ba–polymer complex precursor played a key role in the autocatalytic precipitation reaction and happened in a reaction–diffusion system, resulting in the spontaneous formation of micrometer-sized periodic rings of nanocrystalline $BaCO_3$ grown on the substrate in an aqueous solution [158]. The distance between adjacent rings is almost constant (ca. 5 µm) (Fig. 9a). The numerical simulations using a Brüsselator model for the reaction–diffusion equations qualitatively fit the observed oscillating precipitation reaction well (Fig. 9b). This amazing pattern formation underlies the fact that it is possible to form a spontaneously self-organized pattern in solution by a mesoscale transformation process, which is similar to the patterns observed in a variety of physical, chemical, and natural systems.

3.2.4
Dendrimers

Previously, dendrimers were used as organic matrices for the synthesis of a variety of inorganic nanomaterials and inorganic–organic composites [36, 159]. Dendrimers with different generations have been discovered as active additives for the controlled crystallization of $CaCO_3$ [159, 160]. It has been demonstrated that anionic starburst dendrimers can stabilize spherical vaterite particles for up to a week with controllable particle size in the range of 2.3–5.5 nm in dependence on the dendrimer generation number [161]. In addition, the combination of poly(propylene imine) dendrimers with octadecylamine can stabilize kinetically formed amorphous calcium carbonate (ACC) for periods exceeding 2 weeks in water [162, 163].

3.2.5
Foldamers

Recently, a simple oligopyridine foldamer was designed to recognize the surface of calcite through three carboxylates, projected from one face of the molecule [164]. At low concentrations of the trimer, elongated calcite crystals with angular, toothlike growth, identified as $\{\bar{1}01\}$ faces, were exclusively formed. The ordered array of carboxylates in the foldamer structure exerts a strong influence on the shape of growing calcite crystals via a specific interaction between the foldamer and the newly expressed faces of the growing calcite crystals.

3.2.6
Supramolecular Functional Polymer

Supramolecular directed self-assembly of inorganic and inorganic–organic hybrid nanostructures has emerged as an active area of recent research. The recent advance shows a remarkable feasibility to mimic natural mineralization systems by a designed artificial organic template, where a supramolecular functional polymer can be directly employed as mineralization template for the synthesis of novel inorganic nanoarchitectures [165] such as CdS helices [166] and hydroxyapatite (HAP) nanofibers [167].

4
Synergistic Effects of Crystal Growth Modifiers

4.1
Combination of Mixed Polymer

Usually, mineralization of $BaSO_4$ and $BaCrO_4$ minerals in the presence of poly(ethylene oxide)-*block*-poly(methacrylic acid) (PEO-*b*-PMAA) and a partially monophosphonated derivative, PEO-*b*-PMAA-PO_3H_2 (1%), can produce spherical/elongated particles and fiber bundles/cones [139, 140], respectively (Fig. 10a,b). Recently, a mixture of two DHBCs has been used to control the crystallization and organization of $BaSO_4$ microstructures as a simple model system for the synergistic action of multiple polymers in biomineralization processes [168]. The combinations of the two DHBCs at various weight ratios produced modified forms of these complex morphologies. Nucleation and outgrowth of fiber bundles/cones from spherical precursor particles could be controlled by the polymer mixing ratio to produce materials

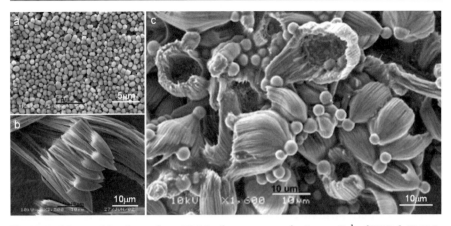

Fig. 10 BaSO$_4$ particles formed at pH 5 in the presence of **a** 1 mg mL^{-1} of PEO-b-PMAA, **b** 2 mg mL^{-1} of PEO-b-PMAA-PO$_3$H$_2$ (1%), and **c** 7:3 w/w mixtures of PEO-b-PMAA-PO$_3$H$_2$ (1%) and PEO-b-PMAA. (Reproduced from [168], © 2004, RSC)

with a shuttlecock-like microstructure (Fig. 10c). Disk-based cones, banded cones, or interconnecting sheets of coaligned fiber bundles can be produced at a total polymer concentration of 3 mg mL^{-1} by a cooperative mechanism involving the combined interaction of both DHBCs with growing BaSO$_4$ crystals [168]. The results reveal that the use of polymer mixtures as additives can provide new variables for crystal morphogenesis compared to systems involving an individual polymer.

4.2
Combination of Polymer with Low Mass Surfactant Molecules

The combination of polymer with low molecular mass surfactant molecules can achieve new synergistic effects on the controlled growth of inorganic crystals. Hollow structures of calcite and disklike hollow vaterite particles can be obtained by the cooperative template effects of the complex micelles formed by PEO-b-PMAA and sodium dodecyl sulfate (SDS) and remaining free DHBC as inhibitor in solution [169]. Similarly, the cationic surfactant cetyltrimethylammonium bromide (CTAB), which can complex the anionic PMAA groups of the DHBC, is able to induce the formation of unusual calcite pinecone-shaped particles. This concept can be further extended to synthesize hollow sub-micrometer sized Ag spheres [170]. Recently, two different soluble polymers (PEG and PMAA) were shown to cooperate with a traditional surfactant (SDS) for the synthesis of spherical calcium carbonate assemblies (e.g., hollow spheres), for which both the morphology and polymorph of the produced CaCO$_3$ crystals were controlled by varying the polymer concentration to change the corresponding transformation of the micelle structure [171].

The combination of crystallization control by DHBCs and self-organization of surfactants in an aqueous environment can lead to remarkable new crystalline structures, as also reported for similar but more simple polyelectrolyte/surfactant additives [172, 173]. Elegant featherlike $BaWO_4$ [174] and $BaMoO_4$ [175] nanostructures were prepared under mild conditions in a multistep growth mechanism by a combination of both catanionic reverse micelles (undecyl acid and decylamine) and the block copolymer PEO-b-PMAA itself. Numerous, nearly parallel, single crystalline barbs stand perpendicular on both sides of a polycrystalline central shaft, showing that special template effects can be achieved in a limited experimental window. These structures could be varied from starlike structures to a single shaft by simple variation of the DHBC concentration, although the role of the DHBC in the generation of these complex structures remained unclear.

4.3
Crystallization in a Mixture of Solvents

Mineralization reactions in alcohol, ethanol, isopropanol, and diethylene glycol have rarely been explored [35, 176, 177], and only $CaCO_3$ crystals with morphology such as elongated spheres or inhomogeneous aggregated structures can be obtained. In recent years, several research groups have occasionally focused on the use of different solvent media to control the crystal growth of $CaCO_3$ and other compounds [178–182]. When the crystallization conditions of $CaCO_3$ in the presence of a DHBC are modified by applying water/alcohol solvent mixtures with varying solvent compositions, and thus solvent quality changes for the block copolymer and $CaCO_3$, $CaCO_3$ nanoparticles aggregate with elongated or spherical morphologies [183].

Vaterite microspheres can be crystallized in water solution using starburst dendrimers [184] and poly(ethylene glycol)-b-poly(L-glutamic acid) (PEG(m)-b-pGlu(n)) [185] as crystal growth modifier. However, the vaterite spheres obtained are not of monodisperse feature. Recently, highly monodisperse vaterite microspheres were produced by taking advantage of the synergic effects of the block copolymer and a selectively mixed solvent under control of an artificial DHBC, PEG(m)-b-pGlu(n) in a mixed solvent made of a suitable volume ratio of N,N-dimethylformamide (DMF)/water (Fig. 11) [186]. This mineralization reaction in a mixed sovent may open a new general route for crystallization of minerals with high quality and structural specialty. Obviously, the property of the mixed solvent plays a key role in controlling the growth, polymorphism, and shape of the $CaCO_3$ mineral [186]. As worsening of the solvent quality for DHBCs is against the concept of well-soluble polymer additives, the mineralization in a mixed solvent can result in the formation of externally triggered DHBC aggregation, which will provide new additional experimental variables, as cation-induced micellization has shown. In addition, the mineral solubility product is sim-

Fig. 11 *Top*: The structure of PEG (110)-*b*-pGlu(6), $m = 110$, $n = 6$. *Bottom*: Highly monodisperse vaterite CaCO$_3$ microspheres mineralized in the presence of PEG (110)-*b*-pGlu(6). The volume ratio of ethanol/water is 1 : 1.4. [PEG-*b*-pGlu] = 1 g L^{-1}; 0.6 mL CaCl$_2$ solution (0.1 M) was added to 6 mL mixed solution with different volume ratio. The crystallization reaction proceeded for 7 days at ambient temperature. (Reproduced from [186], © 2006, Wiley)

ultaneously changed in a mixed solvent system, and interesting cooperative morphogenesis scenarios may be achieved.

5
Artificial Interfaces and Matrices for Crystallization

5.1
Monolayers as Interfaces

Monolayers provide a two-dimensional matrix to mimic the biomineralization process for growing inorganic thin films or crystals [187]. The main concept behind this approach is the view pioneered by Lowenstam [188] that protein layers like, for example, the β-sheets in nacre, have an epitaxial arrangement of functional groups to specific crystal faces. The influence of surfactant headgroups with different functionalities on the crystallization of CaCO$_3$ [189–191] and BaSO$_4$ [192–194], under control of monolayers which are formed from saturated long alkyl chain carboxylates, sulfates, amines, and alcohols, has been studied.

Growth of semiconductor nanocrystals under arachidic acid (AA) monolayers by epitaxial matching of the crystals faces and the headgroups of the surfactants leads to the formation of well-shaped nanocrystals exhibiting specific faces [195]. The PbS thin films contained uniform equilateral triangular PbS crystals, which were obtained by exposing the solution to an AA monolayer in a sealed system (Fig. 12, left). The size of the crystals can be varied by controlling the reaction time [195]. The perfect orientation growth from the {111} basal planes can be well explained by matching the AA monolayer and the cubic PbS structures, as illustrated in Fig. 12 (right). The epitaxial growth of PbS from the {111} face resulted from the geometrical complementarity between the monolayer and the {111} face. The Pb – Pb and S – S interionic distances of 4.2 Å in the PbS {111} plane geometrically matched the d{111} spacing of 4.16 Å for the AA monolayer, as shown in Fig. 12 (right) [195]. The size and preferential orientation of PbS nanocrystals could be controlled by doping the AA monolayers with octadecylamine (ODA) [196]. Similarly, crystallization of CdS under an AA monolayer generated rodlike CdS nanostructures [197].

An interesting result that the monolayer of amphiphilic tricarboxyphenylporphyrin iron(III) μ-oxo dimers can produce highly patterned excavations resulting in "chiral" calcite crystals (Fig. 13) was demonstrated by Lahiri et al. [198]. The porphyrin presents a complex semirigid surface array of carboxylate groups intermediate between protein matrices and simple molecules. The monolayers of the μ-oxo iron(III) porphyrin dimer **1** [199] were formed at the air/water interface using a solution of **1** in 3 : 1 chloroform/methanol (1 mg/mL) to spread the film. Strikingly, about 20% of

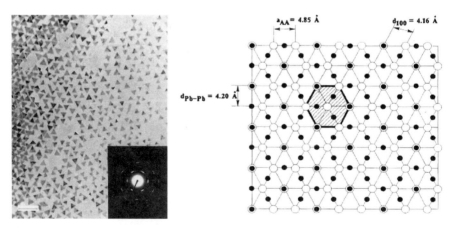

Fig. 12 *Left*: TEM image of a PbS particulate film. The *scale bar* represents 200 nm. The film was obtained by the infusion of H_2S into an arachidic acid (AA) monolayer, floating on an aqueous 5.0×10^{-4} M $Pb(NO_3)_2$ solution in a circular trough, for 45 min. The PbS film was deposited on an amorphous-carbon-coated copper grid. *Right*: Schematic diagram showing the match of the proposed overlap between Pb^{2+} ions and AA headgroups; ○ is AA headgroup and • is Pb^{2+} ions. The *dotted line area* is a unit cell. (Reproduced from [195], © 1995, Wiley)

the three symmetry-related distal {10.4} calcite faces contained impressive rectangular cavities, and some had two or three cavities on adjoining distal {10.4} faces within which one can see clearly the layered and terraced galleries, as depicted in Fig. 13. Significantly, the crystals with a layered excavation on one {10.4} face had rectangular projections on the other, suggesting different views of the same internal structure. The results clearly show that these excavations produce an intrinsically chiral morphology even though calcite has a nonenantiomorphic crystal lattice [198]. The observed "enantiomorphs" of the crystal with chirality could only have originated from the porphyrin dimer template 1, which must have staggered but asymmetric porphyrin planes according to molecular modeling. The molecules were found to

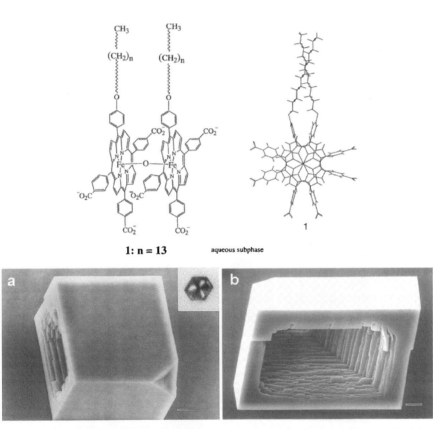

Fig. 13 *Top*: Molecular structure of the μ-oxo iron(III) porphyrin dimer. *Bottom*: SEM images of calcite crystals obtained from nucleation under 1 (**a,b**). **a** Image of a single calcite crystal showing the truncated corner and cavity on a distal {10.4} plane, obtained by the "dipping" method. *Inset*: An in situ optical micrograph of a calcite crystal nucleated under 1, viewed from above. **b** View of the rectangular projections and a terraced gallery on adjoining {10.4} planes distal to the truncated corner. The *scale bar* represents 5 μm. (Reproduced from [198], © 1997, American Chemical Society)

be incorporated inside the crystal by adsorption of porphyrin from accessible surfaces, which confirmed that the highly textured laminations could be due to the anisotropic adsorption of the template 1 on specific planes of calcite crystals.

Recently, Aizenberg et al. showed an impressive example to realize face-selective nucleation of calcite on self-assembled monolayers (SAMs) of alkanethiols in which only the orientation of the functional group is varied [200]. The odd- (C_7, C_{11}, and C_{15}) and even-length (C_{10}, C_{14}, and C_{16}) carboxylic acid terminated alkylthiols supported on silver induced the nucleation of calcite from the (012) face on the basis of computer simulations on the orientation of crystals observed by SEM (Fig. 14a,b). In contrast, SAMs on gold can induced the highly oriented formation of calcite in two distinct crystallographic directions with, respectively, calcite growth from a range of (01l) faces (l = 2–5) (Fig. 14c) for the odd-length alkylthiols (C_7, C_{11}, and C_{15}), and from the (11l) crystallographic planes for the even chain length thiols (C_{10}, C_{14}, and C_{16}) (l = ca. 3, Fig. 14d). The results show that the variation of the orientation of the terminal groups of the molecular template could also regulate the oriented growth of crystals besides controlling the functionality and the lattice of the templating surface [200]. The lateral alignment of {012} habit-modified calcite crystals with respect to a carboxylic acid SAM of thiols on Au(111) substrate in a Kitano solution (pH 5.6–6.0) has been reported [201], implying that it is possible to precisely control the nucleation and grow other inorganic crystals with a preferred orientation to a SAM.

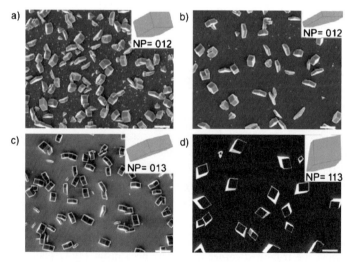

Fig. 14 SEM micrographs of calcite crystals grown on carboxylate-functionalized self-assembled monolayers of **a** C_{15} – Ag, **b** C_{10} – Ag, **c** C_{15} – Au, and **d** C_{10} – Au. The *scale bar* represents 20 µm. *Insets*: Computer simulations of similarly oriented calcite rhombohedra with the nucleating planes (NP) indicated. (Reproduced from [200], © 2003, Wiley)

In addition, crystal growth in the presence of an additive (Mg ions) can be coupled with control over the oriented nucleation achieved by using SAMs as nucleation templates [202].

5.2
Biopolymer Matrix

In Nature, there exist a few biological structures with sophisticated arrangements, such as bacterial threads [203], echinoid skeletal plates [204], eggshell membranes [205], insect wings [206], pollen grains [207], plant leaves [208], wood [209], bacterial cellulose membranes [210], filter paper, cloth, and cotton [211, 212], which were chosen as hard templates to prepare mesoporous inorganic materials with specific structures. However, heat treatment at high temperature is required to remove organic templates from the inorganic–organic hybrid materials. From the viewpoint of biomimetics, various biopolymer matrices can be chosen as templates for the synthesis of a rich family of functional materials with unusual shapes and structures. The biopolymer matrices used for material synthesis include viruses and bio-gels.

Tobacco mosaic virus (TMV) is a very stable tubelike structure of a helical RNA composed of ~ 6400 bases and 2130 identical coat proteins [213, 214], which can be used as a template in the synthesis of nickel and cobalt nanowires [215], cocrystallization of CdS and PbS, iron oxides, and silica [216]. In addition, TMV particles tend to form nematic liquid crystals at a high concentration, which was used to replicate the meso nematic structure for producing mesostructured silica with periodicities of about 20 nm [217]. It is well known that peptides have limited ability of controlling composition, size, and phase during nanoparticle nucleating. Peptide of A7 and J140 has been successfully used as a hard template to prepare CdS and ZnS nanowires [218, 219], and selected M13 bacteriophage to obtain ZnS quantum dot–virus hybrid materials, single-crystal semiconductor (ZnS, CdS) [220], and magnetic (CoPt and FePt) nanowires [221].

Bio-gels are denatured protein with high molecular weight. Generally, they are hydrophilic with various functional groups and different physical and chemical properties from traditional reaction media, for example, water. The reduction of the apparent diffusion rate of the solutes in gels would decrease the nucleation and induced crystal growth through a diffusion-limited process. Theoretical studies have shown that a decrease of diffusivity can lead to a transformation process from an anisotropic shape of diffusion-limited aggregate (DLA) into an irregularly branching pattern [222]. Crystallization of fluorapatite aggregates in a gelatine gel results in the formation of a dumbbell-sphere fractal growth feature, where the intrinsic dipole electric fields may be answered for such a growth mode [223, 224]. It has been recently demonstrated that there is direct correlation between intrinsic fields caused by a parallel orientation of triple-helical protein fibers of gelatine and the self-organized growth of the biocomposite system fluorapatite–gelatine based on an electron holography imaging technique [225].

It is noted that the surface of the sphere is composed of closely packed needlelike units; this kind of crystal is called a "mesocrystal" [226]. Mesocrystals often had a higher symmetry than their constituent tectons [226]. The reactions in gels which are expected to generate mesocrystals need very high supersaturation, leading to increased nucleation number of small clusters as building units for assembly to mesocrystals [227]. Porous hexagonal prism single-crystalline $CaCO_3$ crystals were obtained via self-oriented attachment of the nanocrystals from a reaction between urea and calcium nitrate in a lime-cured gelatine system [228].

Imai and coworkers extended the research of crystal growth in gel media. Their results [229, 230] indicated that in agar, gelatine, and pectin gels, crystals of triclinic systems (H_3BO_3 and $K_2Cr_2O_7$) tended to form peculiar curved and helical branches because of the lowest symmetry of subunits in the polycrystalline material. In the diffusion field (gel system), the connected joints of twinned crystals in aggregates deviated from each other. Unique DLA-like morphologies including twisted branches can be obtained at a band of further increase in gel density, while cubic crystals (NH_4Cl and $Ba(NO_3)_2$) were never found in twisted form under any conditions.

Regular surface-relief calcium carbonate structures were prepared by a cooperated directing agent of soluble poly(acrylic acid) (PAA, $M_w = 2000$) and a substrate of cholesterol-bearing pullulans (Fig. 15) [231]. At low temperature, adapted PAA concentration, and with cholesteryl groups, a high quality periodic calcite structure can be formed by a self-organization process in the reaction–diffusion system with competition between precipitation and ion diffusion.

A predesigned, self-assembled organogelator as template, which resembles the organic matrix in biomineralization used by organic systems to transcribe

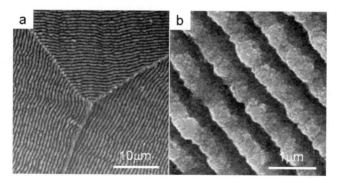

Fig. 15 SEM images of patterned $CaCO_3$ crystals grown on CHP-3 matrices, 20°, 2 days, [PAA] = 2.4×10^{-3} wt %. **a** A boundary among patterned films; **b** a magnified image. (Reproduced from [231], © 2003, Wiley)

inorganic nanostructures, has been intensively studied [18]. A recent paper first reported the synthesis of the nacre morphology by a remineralization process with the introduction of poly(aspartic acid) on the insoluble organic matrix of *Haliotis laevigata*, as for mineralization in Nature [232].

5.3
Synthetic Polymer Matrix

Imai and coworkers [229, 230] also investigated the crystal growth of the triclinic system (H_3BO_3 and $K_2Cr_2O_7$) and cubic crystals (NH_4Cl and $Ba(NO_3)_2$) in synthetic gels, for example, poly(vinyl alcohol) (PVA, M_w = 22 000) and poly(acrylic acid) (PAA, M_w = 250 000, 35 wt % aqueous solution). They obtained similar results to those that occurred in bio-gels, except there existed almost the same amounts of left- and right-handed helices in PAA and PVA matrices, whereas right-handed helices dominated in the bio-gels such as agar, gelatine, and pectin (Fig. 16). Precise control of the chirality of $K_2Cr_2O_7$ crystals in PAA gel [233] can be achieved by addition of a specified amount of chiral molecules (D- and L-glutamic) which can selectively adsorbed on (010) and ($0\bar{1}0$) faces of $K_2Cr_2O_7$. For a crystal with a relatively high symmetry, orthorhombic K_2SO_4 can also form a helical morphology at a high concentration of PAA under control of a DLA process [234].

Potassium sulfate and potassium hydrogen phthalate both formed a nacre-like structure when the PAA concentration was adjusted to a suitable value [235–237] (Fig. 17). Just like the nacre in Nature (Japanese pearl oyster: *Pinctada fucata*) [235], the nacre-like structure of potassium sulfate–PAA and potassium hydrogen phthalate–PAA can absorb and store dyes [235–237] (Fig. 17). With a different concentration of PAA, the K_2SO_4 crystal exhibited a different hierarchical architecture by "iso-oriented assembly" or "oriented attachment" mechanisms [236].

There also existed another use of synthetic polymers besides synthetic gels as the hard template to influence crystal growth discussed above. In this case, solid synthetic polymers were used as a "real hard template". It has been demonstrated that calcium carbonate favored the formation of the vaterite phase on the poly(vinyl chloride-*co*-vinyl acetate-*co*-maleic acid) substrate in the supersaturated solution prepared from calcium nitrate and sodium dicarbonate solutions at pH 8.50 [238]. Commercial polymer fiber (Nylon 66 and Kevlar 29) can induce crystallization of calcite in solution, but the vaterite phase tends to crystallize on the surface of polymers in the presence of soluble polymer (PVA), and aragonite favors forming on the surface of polymers modified with acid or alkali accompanying PVA [239].

Recently, a new type of synthetic polymer which consists of condensed particles (hard part) and polymeric functional blocks (soft part) has been used for the mineralization of calcium carbonate [240, 241]. Poly(diethylaminoethyl methacrylate)-*b*-poly(*N*-isopropyl acrylamide)-*b*-poly(metha-

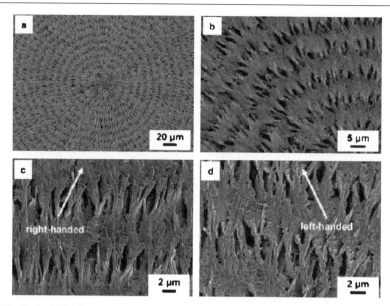

Fig. 16 Typical field-emission SEM (FESEM) images of the $K_2Cr_2O_7$ helical architectures grown in PAA gel matrix without additives after evaporation of water. **a** Spherulitic morphology. **b** Branches having a helical architecture in the spherulite. **c,d** Right- and left-handed helical forms, respectively. (Reproduced from [233], © 2004, American Chemical Society)

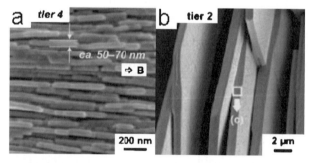

Fig. 17 Typical FESEM images. **a** K_2SO_4 in PAA gel, $C_{PAA} = 8\,\mathrm{g\,dm^{-3}}$. (Reproduced with permission from [236], © 2005, Wiley). **b** Potassium hydrogen phthalate in PAA gel, $C_{PAA} = 10-15\,\mathrm{g\,dm^{-3}}$. (Reproduced from [237], © 2005, RSC)

crylic acid) (PDEAEMA-*b*-PNIPAM-*b*-PMAA; hydrodynamic diameter of the particle in water is about 1 μm), which has both an outer positive (PDEAEMA) and an inner negative (PMAA) block, while the PNIPAM block is regarded as mediating sufficient steric stability, is successfully used to produce stable "sheaf bundle" aragonite with a 1-μm hole in the bottom of the structure (Fig. 18a,b) [240]. A porous single calcite crystal was synthesized

Fig. 18 Typical FESEM images of **a,b** "sheaf bundle" aragonite (reproduced from [240], © 2005, Wiley), and **c,d** porous single calcite (reproduced from [241], © 2004, American Chemical Society)

by poly(styrene(St)-b-methyl methacrylate (MMA)-b-acrylic acid (AA)) with different diameters (Fig. 18c,d) [241]. Crystal growth in the presence of this type of polymer showed remarkably different effects from other polymers and deserves further investigation.

5.4
Crystallization on Foreign External Templates

From the above examples, it is clear that DHBCs themselves can promote the formation of complex crystal morphologies, often involving nanoscopic building units. It is interesting to investigate the influence of an external template on a DHBC-controlled crystallization system. These foreign external templates include CO_2 or air bubbles, charged particles, and air/water interfaces.

In one reported example, the chosen template was as simple as CO_2 bubbles, which are generated by CO_2 evaporation from supersaturated $Ca(HCO_3)_2$ solutions (Kitano method). The gas bubbles were used to act as a foreign external template to produce $CaCO_3$ particles with complex morphologies [134, 137]. Similarly, a combination of the mineralization under control of low molecular weight polyelectrolytes and a foreign static template such as air bubbles has been explored for the generation of macroporous BaCO$_3$ spherulites [93], which is similar to that found in the microemulsion system reported by Mann et al. [242].

Furthermore, charged particles can also be used as templates instead of gas bubbles to break the previous constraint of bundle formation in BaSO$_4$/BaCrO$_4$ crystallization controlled by a phosphonated DHBC [95, 101]. Separated BaCrO$_4$ single crystalline nanofibers with extremely high aspect ratios of > 5000 can be produced through a combination of crystal growth control by DHBCs and controlled nucleation provided by cationic colloidal particles [143].

It has been proved that the basic polymers, which will be positively charged under acidic conditions, are also useful templates for the growth of calcium carbonate [243]. Hemispherical vaterite and needlelike aragonite can be selectively synthesized at the air/water interface by the mediation of poly(ethylene imines)(PEIs) dissolved in supersaturated calcium bicarbonate solution with different molecular weight of PEI blocks, suggesting that cationic polychains are also versatile templates for artificial material synthesis.

Recently, the influence of poly(ethylene glycol)-*block*-poly(ethylene imine) (PEG-*b*-PEI-linear), which is a family of cationic DHBCs, on the crystallization of calcium carbonate at the air/water interface has been studied [244]. The results demonstrated that the crystal morphology of calcium carbonate with layered structures formed at the air/water surface can be well controlled with different PEI block lengths and pH values of the initial solution. The results demonstrated that either PEI length or the solution acidity has significant influence on the morphogenesis of vaterite crystals at the air/water interface (Fig. 19). A possible mechanism for the stratification of CaCO$_3$ vaterite crystals has been proposed. Increasing either PEI length or the initial pH value of the solution will decrease the density of the PEG block anchored

Fig. 19 SEM images of CaCO$_3$ superstructures formed at an air/water interface in the presence of PEG$_{5000}$-*b*-PEI$_{1200}$; the initial pH value is 4. **a** PEG$_{5000}$-*b*-PEI$_{1200}$, 2 g L^{-1}; **b**, **c** PEG$_{5000}$-*b*-PEI$_{400}$; the initial pH values are 4 and 3, respectively. The mineralization time is 5 days. (Reproduced from [244], © 2006, American Chemical Society)

on the binding interface and result in exposing more space as binding interface to solution and favoring the subnucleation and stratification growth on the polymer/CaCO$_3$ interface. In contrast, a higher density of PEG blocks will stabilize the growing crystals more efficiently and inhibit subnucleation on the polymer/CaCO$_3$ interface, thus preventing the formation of stratified structures. This study provides an example which shows that it is possible to access the morphogenesis of calcium carbonate structures by the combination of a block copolymer with an air/water interface.

The above results emphasized that the foreign external templates can be combined with a polymer or polyelectrolyte and provide an additional tool for controlled morphogenesis of inorganic minerals.

5.5
Crystallization on Patterned Surfaces

Crystallization on the monolayer interface and on external templates has been intensively explored in the past few decades (Sects. 5.1 and 5.4). Emerging trends in the field of surface induced crystallization have been focused on how to precisely control the crystallization events, the density and pattern of nucleation events, and the sizes and orientations of the growing crystals. Micropatterned self-assembled monolayers afford control over all these parameters as demonstrated in the crystallization of inorganic crystals [245] and organic crystals [246]. Aizenberg reviewed the bio-inspired approaches to artificial crystallization based on the principles of biomineralization occurring within specific microenvironments [17]. Tailoring of self-assembled monolayers (SAMs) based on modern soft lithography techniques can provide well-defined patterned surfaces for the nucleation of calcium carbonate. Crystallization on these patterned surfaces can result in the formation of large-area, high-resolution inorganic replicas of the underlying organic patterns with advantages including controlled localization of particles, nucleation density, crystal sizes, crystallographic orientation, morphology, polymorphism, stability, and architecture [17, 245, 247].

Biological systems provide numerous examples of micropatterned inorganic materials that directly develop into their intricate architectures, as illustrated by skeleton formation in echinoderms with component function of specialized photosensory organs [19, 248]. Each skeletal structural unit (spines, test plates) is composed of a single calcite crystal delicately patterned on the micrometer scale, which is composed of a close-set array of hemispherical calcitic structures (40–50 µm in diameter) with a characteristic double-lens design (Fig. 20).

In order to mimic such biological structure found in echinoderms, Aizenberg et al. developed a bio-inspired approach to growing large micropatterned calcitic single crystals with controlled orientation and microstructure by crystallization on well-designed micropatterned substrates [247], repre-

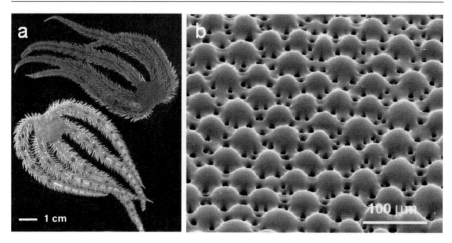

Fig. 20 a The same individual of the brittlestar *Ophiocoma Wendtii*, photographed during the day (*top*) and during the night (*bottom*). **b** SEM image of an array of microlenses on the surface of the dorsal arm plate in *O. Wendtii*. (Reproduced from [248], © 2004, RSC)

senting a general strategy for the design of micro- and nanopatterned crystalline materials (Fig. 21a). In this approach, micropatterned templates, which were organically modified to induce the formation of metastable amorphous calcium carbonate, were nicely imprinted with calcite nucleation sites, resulting in successful template-directed deposition and crystallization of the

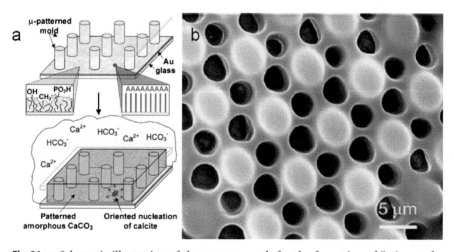

Fig. 21 a Schematic illustration of the new approach for the formation of "microperforated" single crystals: deposition of the ACC mesh from the $CaCO_3$ solution, oriented nucleation at the imprinted nucleation site, and the amorphous-to-crystalline transition of the ACC film on the engineered 3D templates. (Reproduced from [247], © 2003, American Association for the Advancement of Science)

amorphous phase into millimeter-sized single calcite crystals with sub-10-μm patterns and controlled crystallographic orientation (Fig. 21b), which quite resembles the natural skeleton in echinoderms (Fig. 20b). The results indicated that the predesigned three-dimensional templates not only stabilize the ACC and control the oriented crystal nucleation and the micropattern of single crystals, but also act as stress release sites and discharge sumps for excess water and impurities during crystallization [246]. Recently, three-beam interference lithography was used to create a synthetic, biomimetic analog of the brittlestar microlens array with integrated lenses and pores [249].

Such biomimetic synthetic microlens arrays could be potentially used as highly tunable optical elements for a wide variety of applications [246, 247, 249]. The successful fabrication of micropatterned single crystals resembling the natural echinoderm calcitic structures demonstrated that the inspiration from Nature's methods of biological manufacture is proving to be a rich reservoir for the fabrication of advanced materials and devices with novel and superior properties.

6
Summary and Outlook

In summary, recent progress on bio-inspired crystal growth by synthetic templates has been overviewed. Soluble polymer soft templates such as biopolymers and synthetic polymers have shown remarkable effects on the directed crystal growth and controlled self-assembly of inorganic nanoparticles. Different morphogenesis mechanisms of crystal growth such as selective adsorption, mesoscopic transformations, and higher order assembly have been discussed. Recent new developments demonstrated that these soluble polymer soft templates can be combined with other low mass organic molecules or be used in a mixed solvent system to achieve flexible synergistic effects on the mineralization and controlled growth of inorganic crystals with complex form. In contrast, insoluble polymers with different functionalities can be used as a hard template or substrates that offer suitable crystallization sites for the guided crystallization and self-assembly processes. Crystallization on artificial interfaces including monolayers, biopolymer and synthetic polymer matrices for controlled crystal growth, and emerging crystallization on patterned surfaces for the creation of patterned crystals provide additional means to control morphology, microstructure, complexity, and length scales of various inorganic nanostructured materials with two and three dimensionalities.

Recent advances have demonstrated that synthetic template directed crystal growth and mediated self-assembly of nanoparticles can provide promising ways for the rational design of various ordered inorganic and inorganic–organic hybrid materials with complexity and structural specialty. However, there is still a lack of understanding of the nucleation, crystallization, self-assembly, and growth mechanisms of complex superstructures in solution systems. Further multidisciplinary efforts are needed to overcome analytical difficulties which are associated with investigating a multistep and multicomponent morphogenesis mechanism in solution systems. Especially, the detailed interactions on the "soft/hard" interfaces (organic/inorganic interfaces) have to be studied systematically with the help of multianalytical techniques to reveal the real mechanism for the formation of complex matter in solution.

Further exploration in these areas should open new avenues for rationally designing various kinds of inorganic and inorganic–organic hybrid materials with ideal hierarchy at controllable length scales, and desirable dimensionality by bio-inspired approaches. In addition, the study of the relationship between the structural specialty/complexity (shape, size, phase, dimensionality, hierarchy etc.) of the synthetic materials by bio-inspired approaches and their properties will shed new light on the potential but important applications of these materials in various fields in the future.

Acknowledgements S.-H. Yu thanks for the special funding support by the Century Program of the Chinese Academy of Sciences, and the Natural Science Foundation of China (NSFC, Contract Nos. 20325104, 20321101, and 50372065), the Scientific Research Foundation for the Returned Overseas Chinese Scholars supported by the State Education Ministry, the Specialized Research Fund for the Doctoral Program (SRFDP) of Higher Education State Education Ministry, and the Partner Group of the Chinese Academy of Sciences—the Max Planck Society.

References

1. Mann S (1997) In: Bruce DW, O'Hare D (eds) Inorganic materials. Wiley, New York, pp 256–311
2. Mann S, Webb J, Williams RJP (1989) Biomineralization. Wiley, Weinheim
3. Mann S, Ozin GA (1996) Nature 382:313
4. Mann S (2000) Angew Chem Int Ed 39:3393
5. Dujardin E, Mann S (2002) Adv Mater 14:775
6. Estroff LA, Hamilton AD (2001) Chem Mater 13:3227
7. Ozin GA (1997) Acc Chem Res 30:17
8. Ozin GA (2000) Chem Commun 419
9. Kato T, Sugawara A, Hosoda N (2002) Adv Mater 14:869
10. Cölfen H, Mann S (2003) Angew Chem Int Ed 42:2350
11. Mann S (ed) (1996) Biomimetic materials chemistry. Wiley, New York
12. Davis SA, Breulmann M, Rhodes KH, Zhang B, Mann S (2001) Chem Mater 13:3218
13. Antonietti M (2003) Nat Mater 2:9
14. Weiner SW, Addadi L (1997) J Mater Chem 7:689
15. Dabbs DM, Aksay A (2000) Annu Rev Phys Chem 51:601
16. Mann S (1997) J Chem Soc Dalton Trans 3953
17. Aizenberg J (2004) Adv Mater 16:1295
18. Aizenberg J, Hendler G (2004) J Am Chem Soc 126:9271

19. Aizenberg J, Tkachenko A, Weiner S, Addadi L, Hendler G (2001) Nature 412:819
20. van Bommel KJC, Friggeri A, Shinkai S (2003) Angew Chem Int Ed 42:980
21. Matijević E (1993) Chem Mater 5:412
22. Matijević E (1996) Curr Opin Colloid Interface Sci 1:176
23. Antonietti M, Göltner C (1997) Angew Chem Int Ed 36:910
24. Archibald DD, Mann S (1993) Nature 364:430
25. Yang H, Coombs N, Ozin GA (1997) Nature 386:692
26. Li M, Schnablegger H, Mann S (1999) Nature 402:393
27. Peng XG, Manna L, Yang WD, Wickham J, Scher E, Kadavanich A, Alivisatos AP (2000) Nature 404:59
28. Ahmadi TS, Wang ZL, Green TC, Henglein A, El-Sayed MA (1996) Science 272:1924
29. Gibson CP, Putzer K (1995) Science 267:1338
30. Pileni MP, Ninham BW, Gulik-Krzywicki T, Tanori J, Lisiecki I, Filankembo A (1999) Adv Mater 11:1358
31. Park SJ, Kim S, Lee S, Khim ZG, Char K, Hyeon T (2000) J Am Chem Soc 122:8581
32. Adair JH, Suvaci E (2001) Curr Opin Colloid Interface Sci 5:160
33. Bäuerlein E (2003) Angew Chem Int Ed 42:614
34. Niemeyer CM (2001) Angew Chem Int Ed 40:4128
35. Cölfen H (2001) Macromol Rapid Commun 22:219
36. Yu SH, Cölfen H (2004) J Mater Chem 14:2124
37. Cölfen H, Yu SH (2005) MRS Bull 30:727
38. Yu SH, Chen SF (2006) Curr Nanosci 2:81
39. Wulff G (1901) Z. Krystallogr 34:449
40. Buckley HE (1951) Crystal growth. Wiley, New York
41. Mullin IJW (1971) Crystallization. Butterworths, London
42. Chernov AA (1974) J Cryst Growth 24/25:11
43. Kudora T, Irisawa T, Ookawa A (1977) J Cryst Growth 42:41
44. Berg WF (1938) Proc R Soc London Ser A 164:79
45. Bunn CW (1949) Discuss Faraday Soc 5:132
46. Siegfried MJ, Choi KS (2005) Angew Chem Int Ed 44:3218
47. Siegfried MJ, Choi KS (2004) Adv Mater 16:1743
48. Penn RL, Banfield JF (1999) Geochim Cosmochim Acta 63:1549
49. Penn RL, Oskam G, Strathmann TJ, Searson PC, Stone AT, Veblen DR (2001) J Phys Chem B 105:2177
50. Penn RL, Stone AT, Veblen DR (2001) J Phys Chem B 105:4690
51. Penn RL, Banfield JF (1998) Science 281:969
52. Banfield F, Welch SA, Zhang H, Ebert TT, Penn RL (2000) Science 289:751
53. Cölfen H, Antonietti M (2005) Angew Chem Int Ed 44:5576
54. Yu SH (2005) Biomineralized inorganic materials. In: Nalwa HS (ed) Handbook of nanostructured biomaterials and their applications, vol 1. American Scientific, Los Angeles, pp 1–69
55. Antonietti M (2001) Curr Opin Colloid Interface Sci 6:244
56. Hardikar VV, Matijević E (2001) Colloids Surf A 186:23
57. Shen FH, Feng QL, Wang CM (2002) J Cryst Growth 242:239
58. Belcheer AM, Wu HX, Christensen RJ, Hansma PK, Stucky GD, Morse DE (1996) Nature 381:56
59. Feng QL, Pu G, Pei Y, Cui FZ, Li HD, Kim TN (2000) J Cryst Growth 216:459
60. Raz S, Weiner S, Addadi L (2000) Adv Mater 12:38
61. Belcher AM, Wu XH, Christensen RJ, Hansma PK, Stucky GD, Morse DE (1996) Nature 381:56

62. Falini G, Albeck S, Weiner S, Addadi L (1996) Science 271:67
63. DeOliveira DB, Laursen RA (1997) J Am Chem Soc 119:10627
64. Li CM, Botsaris GD, Kaplan DL (2002) Cryst Growth Des 2:387
65. Arias JL, Fernandez MS (2003) Mater Charact 50:189
66. Mirkin CA (2000) Inorg Chem 39:2258
67. Mirkin CA, Letsinger RL, Mucic RC, Storhoff JJ (1996) Nature 382:607
68. Alivisatos AP, Johnson K, Peng X, Wilson TE, Loweth CJ, Bruchez M, Schultz PG (1996) Nature 382:609
69. Dujardin E, Hsin LB, Wang CRC, Mann S (2001) Chem Commun 1264
70. Lee SW, Lee SK, Belcher AM (2003) Adv Mater 15:689
71. Dujardin E, Peet C, Stubbs G, Culver JN, Mann S (2003) Nano Lett 3:413
72. Mao C, Flynn CE, Hayhurst A, Sweeney R, Qi J, Georgiou G, Iverson B, Belcher AM (2003) Proc Natl Acad Sci USA 100:6946
73. Vali H, Weiss B, Li YL, Sears SK, Kim SS, Kirschvink JL, Zhang CL (2004) Proc Natl Acad Sci USA 101:16121
74. Bharde A, Wani A, Shouche Y, Joy PA, Prasad BLV, Sastry M (2005) J Am Chem Soc 127:9326
75. Mann S, Sparks NHC, Frankel RB, Bazylinskida DA, Jannasch HW (1990) Nature 343:258
76. Rautaray D, Ahmad A, Sastry M (2003) J Am Chem Soc 125:14656
77. Rautaray D, Ahmad A, Sastry M (2004) J Mater Chem 14:2333
78. Rautaray D, Sanyal A, Adyanthaya SD, Ahmad A, Sastry M (2004) Langmuir 20:6827
79. Ahmad A, Rautaray D, Sastry M (2004) Adv Funct Mater 14:1075
80. Shankar SS, Rai A, Ankamwar B, Singh A, Ahmad A, Sastry M (2004) Nat Mater 3:482
81. Shankar SS, Rai A, Ahmad A, Sastry M (2004) J Colloid Interface Sci 275:496
82. Kröger N, Deutzmann R, Sumper M (1996) Science 286:1129
83. Kröger N, Deutzmann R, Bergsdorf C, Sumper M (2000) Proc Natl Acad Sci USA 97:14133
84. Kröger N, Lorenz S, Brunner E, Sumper M (2002) Science 298:584
85. Sumper M (2002) Science 295:2430
86. Menzel H, Horstmann S, Behrens P, Bärnreuther P, Krueger I, Jahns M (2003) Chem Commun 2994
87. Cha JN, Stucky GD, Morse DE (2000) Nature 403:289
88. Cha JN, Shimizu K, Zhou Y, Christiansen SC, Chmelka BF, Stucky GD, Morse DE (1996) Proc Natl Acad Sci USA 96:361
89. Rieger J (2002) Tenside Surfactants Deterg 39:221
90. Peytcheva A, Cölfen H, Antonietti M (2002) Colloid Polym Sci 280:218
91. Gower LB, Tirrell DA (1998) J Cryst Growth 191:153
92. Gower LB, Odom DJ (2000) J Cryst Growth 210:719
93. Yu SH, Cölfen H, Xu AW, Dong WF (2004) Cryst Growth Des 4:33
94. Bigi A, Boanini E, Walsh D, Mann S (2002) Angew Chem Int Ed 41:2163
95. Qi LM, Cölfen H, Antonietti M, Li M, Hopwood JD, Ashley AJ, Mann S (2001) Chem Eur J 7:3526
96. Yu SH, Antonietti M, Cölfen H, Hartmann J (2003) Nano Lett 3:379
97. Olszta MJ, Odom DJ, Douglas EP, Gower LB (2003) Connect Tissue Res 44 (Suppl. 1):326
98. Olszta MJ, Gajjeraman S, Kaufman M, Gower LB (2004) Chem Mater 16:2355
99. Wagner RS, Ellis WC (1964) Appl Phys Lett 4:89
100. Trentler TJ, Hickman KM, Goel SC, Viano AM, Gibbons PC, Buhro WE (1995) Science 270:1791

101. Yu SH, Cölfen H, Antonietti M (2002) Chem Eur J 8:2937
102. Jada J, Verraes A (2003) Colloids Surf A 219:7
103. Shchukin DG, Sukhorukov GB, Mohwald H (2003) Chem Mater 15:3947
104. Wang TX, Cölfen H, Antonietti M (2005) J Am Chem Soc 127:3246
105. Donners JJJM, Nolte RJM, Sommerdijk NAJM (2002) J Am Chem Soc 124:9700
106. Ueyama N, Hosoi T, Yamada Y, Doi M, Okamura T, Nakamura A (1998) Macromolecules 31:7119
107. Ueyama N, Kozuki H, Doi M, Yamada Y, Takahashi K, Onoda A, Okamura T, Yamamoto H (2001) Macromolecules 34:2607
108. Ueyama N, Takeda J, Yamada Y, Onoda A, Okamura T, Nakamura A (1998) Inorg Chem 38:475
109. Sugawara T, Suwa Y, Ohkawa K, Yamamoto Y (2003) Macromol Rapid Commun 24:847
110. Zhang ZP, Gao D, Zhao H, Xie C, Guan G, Wang D, Yu SH (2006) J Phys Chem B 110:8613
111. Xu AW, Antonietti M, Cölfen H, Fang YP (2006) Adv Funct Mater 16:903
112. Xu AW, Yu Q, Dong WF, Antonietti M, Cölfen H (2005) Adv Mater 17:2217
113. Peng Y, Xu AW, Deng B, Antonietti M, Cölfen H (2006) J Phys Chem B 110:2988
114. Kotachi A, Miura T, Imai H (2004) Chem Mater 16:3191
115. Wegner G, Baum P, Müller M, Norwig J, Landfester K (2001) Macromol Symp 175:349
116. Basko M, Kubisa P (2002) Macromolecules 35:8948
117. Basko M, Kubisa P (2004) J Polym Sci A Polym Chem 42:1189
118. Förster S, Plantenberg T (2002) Angew Chem Int Ed 41:689
119. Kriesel JW, Sander MS, Tilley TD (2001) Chem Mater 13:3554
120. Bronstein LM, Sidorov SN, Gourkova AY, Valetsky PM, Hartmann J, Breulmann M, Cölfen H, Antonietti M (1998) Inorg Chim Acta 280:348
121. Sidorov SN, Bronstein LM, Valetsky PM, Hartmann J, Cölfen H, Schnablegger H, Antonietti M (1999) J Colloid Interface Sci 212:197
122. Bronstein LM, Sidorov SN, Valetsky PM, Hartmann J, Cölfen H, Antonietti M (1999) Langmuir 15:6256
123. Yu SH, Cölfen H, Mastai Y (2004) J Nanosci Nanotechnol 4:291
124. Zhang D, Qi LM, Ma JM, Cheng HM (2001) Chem Mater 13:2753
125. Qi LM, Cölfen H, Antonietti M (2002) Nano Lett 1:61
126. Bouyer F, Gérardin C, Fajula F, Putaux JL, Chopin T (2003) Colloids Surf A 217:179
127. Sedlak M, Antonietti M, Cölfen H (1998) Macromol Chem Phys 199:247
128. Sedlak M, Cölfen H (2001) Macromol Chem Phys 202:587
129. Cölfen H, Antonietti M (1998) Langmuir 14:582
130. Cölfen H, Qi LM (2001) Chem Eur J 7:106
131. Marentette JM, Norwig J, Stockelmann E, Meyer WH, Wegner G (1997) Adv Mater 9:647
132. Norwig J (1997) Mol Cryst Liq Cryst 313:115
133. Yu SH, Cölfen H, Hartmann J, Antonietti M (2002) Adv Funct Mater 12:541
134. Rudloff J, Antonietti M, Cölfen H, Pretula J, Kaluzynski K, Penczek S (2002) Macromol Chem Phys 203:627
135. Kaluzynski K, Pretula J, Lapienis G, Basko M, Bartczak Z, Dworak A, Penczek S (2001) J Polym Sci A Polym Chem 39:955
136. Yu S-H, Cölfen H, Antonietti M (2003) J Phys Chem B 107:7396
137. Rudloff J, Cölfen H (2004) Langmuir 20:991
138. Antonietti M, Breulmann M, Göltner C, Cölfen H, Wong KK, Walsh D, Mann S (1998) Chem Eur J 4:2493

139. Qi LM, Cölfen H, Antonietti M (2000) Angew Chem Int Ed 39:604
140. Qi LM, Cölfen H, Antonietti M (2002) Chem Mater 12:2392
141. Cölfen H, Qi LM, Mastai Y, Börger L (2002) Cryst Growth Des 2:191
142. Robinson KL, Weaver JVM, Armes SP, Marti ED, Meldrum FC (2002) J Mater Chem 12:890
143. Yu S-H, Cölfen H, Antonietti M (2003) Adv Mater 15:133
144. Bagwell RB, Sindel J, Sigmund W (1999) J Mater Res 14:1844
145. Zhang DB, Qi LM, Ma JM, Cheng HM (2002) Chem Mater 14:2450
146. Öner M, Norwig J, Meyer WH, Wegner G (1998) Chem Mater 10:460
147. Taubert A, Palms D, Weiss O, Piccini MT, Batchelder DN (2002) Chem Mater 14:2594
148. Taubert A, Palms D, Glasser G (2002) Langmuir 18:4488
149. Taubert A, Kübel C, Martin DC (2003) J Phys Chem B 107:2660
150. Yu SH, Antonietti M, Cölfen H, Giersig M (2002) Angew Chem Int Ed 41:2356
151. Mastai Y, Sedlák M, Cölfen H, Antonietti M (2002) Chem Eur J 8:2430
152. Qi LM (2001) J Mater Sci Lett 20:2153
153. Mastai Y, Rudloff J, Cölfen H, Antonietti M (2002) ChemPhysChem 3:119
154. Chen SF, Yu SH, Wang TX, Jiang J, Cölfen H, Hu B, Yu B (2005) Adv Mater 17:1461
155. Cölfen H (2001) Macromol Rapid Commun 22:219
156. Gao YX, Yu SH, Cong H, Jiang J, Xu AW, Dong WF, Cölfen H (2006) J Phys Chem B 110:6432
157. Yu SH, Cölfen H, Tauer K, Antonietti M (2005) Nat Mater 5:51
158. Wang TX, Xu AW, Cölfen H (2006) Angew Chem Int Ed 45:4451
159. Naka K (2003) Top Curr Chem 228:141
160. Naka K, Tanaka Y, Chujo Y, Ito Y (1999) Chem Commun 1931
161. Naka K, Tanaka Y, Chujo Y (2002) Langmuir 18:3655
162. Donners JJJM, Heywood BR, Meijer WM, Nolte RJM, Roman C, Schenning APLHJ, Sommerdijk NAJM (2000) Chem Commun 1937
163. Donners JJJM, Heywood BR, Meijer WM, Nolte RJM, Sommerdijk NAJM (2002) Chem Eur J 8:2561
164. Estroff LA, Incarvito CD, Hamilton AD (2004) J Am Chem Soc 126:2
165. Hartgerink JD, Zubarev ER, Stupp SI (2001) Curr Opin Colloid Interface Sci 5:355
166. Sone ED, Zubarev ER, Stupp SI (2002) Angew Chem Int Ed 41:1705
167. Hartgerink JD, Beniash E, Stupp SI (2001) Science 294:1684
168. Li M, Mann S, Cölfen H (2004) J Mater Chem 14:2269
169. Qi LM, Li J, Ma JM (2000) Adv Mater 14:300
170. Zhang D, Qi LM, Ma JM, Cheng HM (2002) Adv Mater 14:1499
171. Deng SG, Cao JM, Feng J, Guo J, Fang BQ, Zheng MB, Tao J (2005) J Phys Chem B 109:11473
172. Wei H, Shen Q, Zhao Y, Wang D, Xu D (2004) J Cryst Growth 260:511
173. Wei H, Shen Q, Zhao Y, Wang D, Xu D (2004) J Cryst Growth 260:545
174. Shi HT, Qi LM, Ma JM, Cheng HM (2003) J Am Chem Soc 125:3450
175. Shi HT, Qi LM, Ma JM, Wu NZ (2005) Adv Funct Mater 15:442
176. Manoli F, Dalas E (2000) J Cryst Growth 218:359
177. Qi LM, Ma JM (2002) Chem J Chin Univ 23:1595
178. Dickinson SR, Mcgrath KM (2003) J Mater Chem 13:928
179. Falini G, Gazzano M, Ripamonti A (1996) Chem Commun 1037
180. Seo KS, Han C, Wee JH, Park JK, Ahn JW (2005) J Cryst Growth 276:680
181. Chen SF, Yu SH, Yu B (2004) Chem Eur J 10:3050
182. Chen SF, Yu SH, Jiang J, Li FQ, Liu YK (2006) Chem Mater 18:122
183. Qi LM, Li J, Ma JM (2002) Chem J Chin Univ 23:1595

184. Naka K, Tanaka Y, Chujo Y (2002) Langmuir 18:3655
185. Kašparová P, Antonietti M, Cölfen H (2004) Colloids Surf A 250:153
186. Guo XH, Yu SH, Cai GB (2006) Angew Chem Int Ed 45:3977
187. Heywood BR (1996) Template-directed nucleation and growth of inorganic materials. In: Mann S (ed) Biomimetic materials chemistry. Wiley, New York, pp 143–173
188. Lowenstam HA (1981) Science 211:1126
189. Mann S, Heywood BR, Rajam S, Birchall JD (1988) Nature 334:692
190. Rajam S, Heywood BR, Walker JBA, Mann S, Davey RJ, Birchall JD (1991) J Chem Soc Faraday Trans 87:727
191. Heywood BR, Rajam S, Mann S (1991) J Chem Soc Faraday Trans 81:735
192. Heywood BR, Rajam S, Mann S (1992) J Am Chem Soc 114:4681
193. Heywood BR, Mann S (1992) Langmuir 8:1492
194. Heywood BR, Mann S (1992) Adv Mater 4:278
195. Fendler JH, Meldrum FC (1995) Adv Mater 7:607
196. Yang J, Fendler JH (1995) J Phys Chem 99:5505
197. Yang J, Fendler JH (1995) J Phys Chem 99:5500
198. Lahiri J, Xu G, Dabbs DM, Yao N, Aksay IA, Groves JT (1997) J Am Chem Soc 119:5449
199. Lahiri J, Fate GF, Ungashe SB, Groves JT (1996) J Am Chem Soc 118:2347
200. Han YJ, Aizenberg J (2003) Angew Chem Int Ed 42:3668
201. Travaille AM, Kaptijn L, Verwer P, Hulsken B, Elemans JAAW, Nolte RJM, van Kempen H (2003) J Am Chem Soc 125:11571
202. Han YJ, Laura M, Wysocki LM, Thanawala MS, Siegrist T, Aizenberg J (2005) Angew Chem Int Ed 44:2386
203. Davis SA, Burkett SL, Mendelson NH, Mann S (1997) Nature 385:420
204. Meldrum FC, Seshadri R (2000) Chem Commun 29
205. Yang D, Qi LM, Ma JM (2002) Adv Mater 14:1543
206. Cook G, Timms PL, Goeltner-Spickermann C (2003) Angew Chem Int Ed 42:557
207. Hall SR, Bolger H, Mann S (2003) Chem Commun 2784
208. Valtchev V, Smaihi M, Faust AC, Vidal L (2003) Angew Chem Int Ed 42:2782
209. Shin Y, Wang C, Exarhos GJ (2005) Adv Mater 17:73
210. Zhang DY, Qi LM (2005) Chem Commun 2735
211. Huang J, Kunitake T (2003) J Am Chem Soc 125:11834
212. Shin Y, Li XS, Wang C, Coleman JR, Exarhos GJ (2004) Adv Mater 16:1212
213. Falvo MR, Washburn R, Superfine R, Finch M, Brooks FP Jr, Chi V, Taylor RM II (1997) Biophys J 72:1396
214. Watson JD (1954) Biochim Biophys Acta 13:10
215. Knez M, Bittner AM, Boes F, Wege C, Jeske H, Mai E, Kern K (2003) Nano Lett 3:1079
216. Shenton W, Douglas T, Young M, Stubbs G, Mann S (1999) Adv Mater 11:253
217. Fowler CE, Shenton W, Stubbs G, Mann S (2001) Adv Mater 13:1266
218. Whaley SR, English DS, Hu EL, Barbara PF, Belcher AM (2000) Nature 405:665
219. Mao C, Flynn CE, Hayhurst A, Sweeney R, Qi J, Georgiou G, Iverson B, Belcher AM (2003) Proc Natl Acad Sci USA 100:6946
220. Lee SW, Mao CB, Flynn CE, Belcher AM (2002) Science 296:892
221. Mao C, Solis DJ, Reiss BD, Kottmann ST, Sweeney RY, Hayhurst A, Georgiou G, Iverson B, Belcher AM (2004) Science 303:213
222. Bogoyavlenskiy VA, Chernova NA (2000) Phys Rev E 61:1629
223. Busch S (1998) Dissertation, Technische Universität Darmstadt
224. Busch S, Dolhaine A, Duchesne A, Heinz S, Hochrein O, Laeri F, Podebrad O, Vietze U, Weiland T, Kniep R (1999) Eur J Inorg Chem 10:1643

225. Simon P, Zahn D, Lichte H, Kniep R (2006) Angew Chem Int Ed 45:1911
226. Cölfen H, Antonietti M (2005) Angew Chem Int Ed 44:5576
227. Putnis A, Prieto M, Fernandez-Diaz L (1995) Geol Mag 132:1
228. Zhan JH, Lin HP, Mou CY (2003) Adv Mater 15:621
229. Imai H, Oaki Y (2004) Angew Chem Int Ed 43:1363
230. Oaki Y, Imai H (2003) Cryst Growth Des 3:711
231. Sugawara A, Ishii T, Kato T (2003) Angew Chem Int Ed 42:5299
232. Gehrke N, Nassif N, Pinna N, Antonietti M, Gupta HS, Cölfen H (2005) Chem Mater 17:6514
233. Oaki Y, Imai H (2004) J Am Chem Soc 126:9271
234. Oaki Y, Imai H (2005) Langmuir 21:863
235. Oaki Y, Imai H (2005) Angew Chem Int Ed 44:6571
236. Oaki Y, Imai H (2005) Adv Funct Mater 15:1407
237. Oaki Y, Imai H (2005) Chem Commun 6011
238. Dalas E, Klepetsanis P, Koutsoukos PG (1999) Langmuir 15:8322
239. Lakshminarayanan R, Valiyaveettil S, Loy GL (2003) Cryst Growth Des 3:953
240. Nassif N, Gehrke N, Pinna N, Shirshova N, Tauer K, Antonietti M, Cölfen H (2005) Angew Chem Int Ed 44:6004
241. Lu CH, Qi LM, Cong HL, Wang XY, Yang JH, Yang LL, Zhang DY, Ma JM, Cao WX (2005) Chem Mater 17:5218
242. Walsh D, Lebeau B, Mann S (1999) Adv Mater 11:324
243. Park HK, Lee I, Kim K (2004) Chem Commun 24
244. Gao YQ, Yu SH, Cong HP (2006) Langmuir 22:6125
245. Aizenberg J, Black AJ, Whitesides GM (1999) Nature 398:495
246. Briseno AL, Aizenberg J, Han YJ, Penkala RA, Moon H, Lovinger AJ, Kloc C, Bao ZN (2005) J Am Chem Soc 127:12164
247. Aizenberg J, Muller DA, Grazul JL, Hamann DR (2003) Science 299:1205
248. Aizenberg J, Hendler G (2004) J Mater Chem 14:2066
249. Yang S, Chen G, Megens M, Ullal CK, Han YJ, Rapaport R, Thomas EL, Aizenberg J (2005) Adv Mater 17:435

Delayed Action of Synthetic Polymers for Controlled Mineralization of Calcium Carbonate

Kensuke Naka

Department of Polymer Chemistry, Graduate School of Engineering, Kyoto University, Katsura, Nishikyo-ku, 615-8510 Kyoto, Japan
ken@chujo.synchem.kyoto-u.ac.jp

1	Introduction	120
2	Calcium Carbonate	123
2.1	Precipitation Experiments	123
2.2	Amorphous Calcium Carbonate (ACC)	125
2.3	Vaterite	126
2.4	Aragonite	128
2.5	Calcite	128
3	Multi-Step Nucleation and Growth in the Mineralization Process	129
3.1	Natural Organism	129
3.2	In Vitro Study	130
3.3	Matrix-Mediated Formation of $CaCO_3$ Thin Films	131
4	Delayed Action for Nucleation and Growth of $CaCO_3$	132
4.1	Overview of Calcium Carbonate Crystallization by Synthetic Substrates	133
4.2	Delay Addition of Poly(acrylic acid)	133
4.3	In-situ Polymerization	134
4.4	Delay Addition of Anionic Polyamidoamine Dendrimers	139
4.5	Polymers Responsive to Stimulation	143
5	Template Mineralization on the Surface of Calcium Carbonate	147
6	Conclusions and Outlook	150
	References	151

Abstract Natural inorganic–organic hybrid materials are formed through mineralization of inorganic materials on self-assembled organic materials. In these mineralized tissues, crystal morphology, size, and orientation are determined by local conditions and, in particular, the presence of "matrix" proteins or other macromolecules. The final crystalline phase arises through a series of steps initiated by the formation of an amorphous phase that undergoes subsequent phase transformations. The multi-step crystallization process on living systems was supported by the detection of different mineral polymorphs in natural organisms and subsequent phase transformation. This work focuses on a new concept for controlling crystal polymorphs by delayed action of organic additives during nucleation stages. During the formation of continuous thin films of minerals, several authors have used a phase-transformation process from an initially deposited amorphous phase to crystalline phase. The delay addition method gives a new simple process for controlling the $CaCO_3$ crystallization. Three different crystal polymorphs of $CaCO_3$ (aragonite, vaterite, and calcite) were selectively induced by changing the time when the radical initiator was added to a calcium carbonate solution with sodium acrylate. These processes may be similar to the secretion of specific proteins or molecules during the transformation of biomineralization.

Keywords Calcium carbonate · Delay addition · In situ polymerization · Latent inductor · Transformation

Abbreviations
ACC   amorphous calcium carbonate
DMSO dimethyl sulfoxide
FT-IR Fourier transform infrared spectroscopy
h   hour(s)
KPS   potassium peroxodisulfate
L   liter(s)
min   minute(s)
mol   mole(s)
M_w weight average molecular weight
PAA   poly(acrylic acid)
PAZO polyazobenzen
PVP   poly(N-vinylpyrrolidone)
SEM   scanning electron microscopy
TEM   transmission electron microscopy
XRD   X-ray diffraction
UV   ultra violet

1
Introduction

The design of nanomaterials, in other words, hybrid materials, has emerged as one of the most exciting areas of scientific effort in this decade. Among various research fields aimed at constructing nanomaterials, organic–inorganic hybrid materials have opened a new horizon in the field of materials science. When different materials are hybridized at the nano-meter scale, the obtained hybrid materials show unique properties compared with microscale composites [1–3]. Organic–inorganic hybrids have been elaborated with various inorganic hosts such as inorganic clay compounds [4], metal oxo clusters [5], oligosilsesquioxanes and their derivatives [6], zeolite [7], and metal [8] nanoparticles. Among them, sol-gel reaction of metal alkoxides is a widely used technique for the preparation of the organic–inorganic hybrid materials [9–11]. One of the most important advantages of the sol-gel process for preparation of the hybrids is the milder process compared with the normal glass preparation method. The mild characteristics offered by the sol-gel reaction allow the introduction of organic components inside the inorganic network. Organic–inorganic polymer hybrids can be prepared by mixing organic polymer into the sol-gel reaction. To obtain homogeneous organic–inorganic hybrids, in which the organic polymer is dispersed in the inorganic matrix at the nano or molecular scale, increased compatibility between the organic polymer and inorganic phases is necessary. The introduction of covalent bonds or chemical and physical interactions between the organic polymers and the inorganic units are efficient at increasing compatibility.

Nanomaterials are assembled from simpler components such as molecules, polymers, and other nanostructures under mild conditions. This approach is similar to the one nature uses to construct complex biological architecture. In nature, biological organisms produce polymer-inorganic hybrid materials such as bone, teeth, diatoms, and shells. These hybrids have superior mechanical properties as compared to synthetic hybrids. For example, the abalone shell, a composite of calcium carbonate with a few percent of the organic component (Fig. 1), is 3000 times more fracture resistant than a single crystal of the pure mineral [12–14]. The core of the organic template is composed of a layer of β-chitin layered between "silk-like" glycine- and alanine-rich proteins. The outer surfaces of the template are coated with hydrophilic acidic macromolecules (Fig. 1). Natural inorganic–organic hybrid materials are formed through mineralization of inorganic materials on self-assembled organic materials. In these mineralized tissues, crystal morphology, size, and orientation are determined by local conditions and, in particular, the presence of "matrix" proteins or other macromolecules [15]. Biopolymers and low-molecular-weight organic molecules are organized into nanostructures and used as frameworks for specifically oriented and shaped inorganic crystals such as calcium carbonate, hydroxyapatite, iron oxide, and silica. These processes use an aqueous solution at temperature below 100 °C. Although many useful and characteristic organic–inorganic hybrids have been developed, the naturally produced hybrid materials are superior to these artificial materials.

The interest of most researchers lies in understanding how organized inorganic materials with complex morphological forms can be produced by

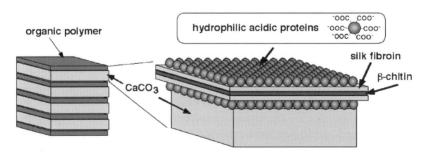

Fig. 1 Schematic illustration of the nacre of the abalone shell

biomineralization processes, and how such complexity can be reproducibly synthesized in biomimetic systems. In vitro studies of biomineralization have provided useful information for the design of organic templates. Falini and co-workers assembled in vitro a complex containing the major matrix components present in a mollusk shell, namely β-chitin, silk-fibroin-like protein, and water-soluble acidic macromolecules [16]. When this assembly was placed in a saturated solution of calcium carbonate, multi-crystalline spherulites formed within the complex. They extracted aspartic acid-rich glycoproteins from an aragonitic mollusk shell layer or a calcitic layer. These were aragonite if the added macromolecules were from the aragonitic shell layer, or calcite if they were derived from the calcitic shell layer. In the absence of the acidic glycoproteins, no mineral was formed within the complex. Because of a lack of information on the structure of the proteins, a detailed mechanism of the effect of nucleation at the molecular level remains unknown. However, in spite of this, these in vitro studies motivate us to design artificial templates for the controlled nucleation of minerals.

Because of the complexity of the natural biomineralization systems, research on mineralization has been carried out on model organic interfaces. At the beginning, these studies were focused on the main fundamental question of how inorganic crystallization can be controlled in an aqueous solution. Construction of organic–inorganic hybrid materials with controlled mineralization is currently of interest to both organic and inorganic chemists to understand the mechanism of natural biomineralization process as well as to seek industrial and technological applications. Although many researchers have realized that the design of organic templates is important for controlled mineralization of inorganic materials, many of the biomineralization studies have been approached from the viewpoint of inorganic materials chemistry. Recently, several organic and polymer research scientists became interested in the research area of biomineralization. Various types of organic matrices as a structural template for crystallization of inorganic materials should be easily designed and synthesized [17, 18].

The majority of these efforts have focused on exploring the characteristic effects of templates on crystal nucleation and growth. On the specific interactions, various synthetic polymers have been found to be potent inhibitors or habit modifiers for inorganic crystallization by adsorption onto the surfaces of the growing crystals, thus controlling their growth rate and habit through the strength and selectivity of this adsorption [19–21]. The final crystalline phase arises through a series of steps. By selectively interacting with the minerals at different stages during the crystal forming process, the organisms may choose manipulating both the polymorph and the orientation of the mineral to meet specific biological requirements. Although the presence of various synthetic additives has been studied for inorganic crystallization, selective interaction of an organic matrix with the mineral at different stages has been unexplored. This work will focus on a new concept for controlling crystal polymorphs of calcium carbonate by delayed action of polymeric additives after starting nucleation of $CaCO_3$.

2
Calcium Carbonate

The main inorganic mineral produced in natural organisms is calcium carbonate. Calcium carbonate is an attractive model mineral for studies in the laboratory, since its crystals are easily characterized and the morphology of $CaCO_3$ has been the subject of control in biomineralization processes. Therefore, this work is focused on calcium carbonate as the inorganic phase. The precipitation of calcium carbonate in aqueous solution is also of great interest for industrial and technological applications. The effect of pH, foreign ions, organic additives, and the degree of supersaturation in aqueous systems have been extensively studied. The particular interest in this system is due to the polymorphism of calcium carbonate. Crystalline calcium carbonate can adopt several different structures, differing in detail of the lattice structure of the crystal. The atomic structure with the lowest lattice energy, and hence the most stable, is calcite. Less stable, and with a slightly different lattice structure, is aragonite. The most unstable crystalline phase is vaterite. A very unstable form of calcium carbonate called "amorphous calcium carbonate" (ACC) is also found in many instances in nature. Kahmi reported a relationship between the temperature at which calcium carbonate was precipitated [22]. At temperatures below 15 °C, calcite with six-fold coordination of calcium was the predominant phase. Vaterite was found to have calcium coordinated to eight oxygen atoms between 20 to 60 °C. Above 60 °C aragonite was coordinated to nine oxygen atoms. Thermal vibrations probably allowed for an increase in the effective radii of the calcium atoms, allowing stabilization of the structure. In the following section, an overview of these polymorphs is presented.

2.1
Precipitation Experiments

Crystallization of calcium carbonate is highly dependent on precipitation conditions. In the classical method, much of the scientific investigation of calcium carbonate formation focused on the seed growth of anhydrous calcite crystals from solutions of low supersaturation [23–26]. A crystal can precipitate via epitaxy directly from the liquid solution so that the nuclei bear the same structure as the final crystal.

Spontaneous precipitation by the mixing of two concentrated solutions of calcium and carbonate results in a gelatinous matter when ionic activity product exceeds the solubility product of amorphous calcium carbonate.

The spontaneous precipitation provides rather high supersaturation. At such high supersaturations, the local conditions of temperature, concentration, and impurities can have a profound effect on the qualitative nature of the phases formed [27]. Under this condition, amorphous calcium carbonate appears as an intermediate phase. This amorphous precursor is highly unstable and transforms into crystalline polymorphs within a few minutes if kept in solution.

One of the promising techniques to control the precipitation process is the double-jet method (Fig. 2) [28]. The two reactants ($CaCl_2$ and Na_2CO_3) are injected via capillaries into a reaction vessel under vigorous stirring to prevent heterogeneous nucleation at the glass wall [29, 30]. The two capillary ends are joined together so that a high local reactant concentration and thus extreme supersaturation is achieved at the moment when the two reactants leave the capillaries, which provides an immediate nucleation of $CaCO_3$. The nuclei are then immediately transported to regions of lower $CaCO_3$ concentration and can grow further. The $CaCO_3$ crystal formation occurring after an excess addition of reactants was easily observed as a sudden increase in the turbidity of the solution. The main idea behind this technique, which was set up for the controlled precipitation of silver halides in the photographic industry, is to maintain a rapid nucleation of a constant particle number at the beginning of the experiment to enable growth of monodispersed particles [28]. Under this precipitation condition, extreme supersaturation is achieved at a particular moment and this provides an immediate nucleation of $CaCO_3$, which is not affected by the organic additives. The nuclei are then immediately transported to the regions of lower $CaCO_3$ concentration and can grow further.

A carbonate diffusion method provides slow crystallization of calcium carbonate (Fig. 3) [30]. A solution of calcium chloride in distilled water is placed

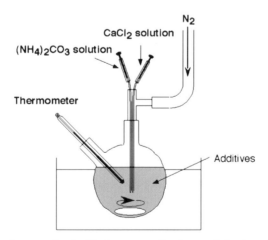

Fig. 2 Experimental setup of a double-jet reactor for the precipitation of $CaCO_3$

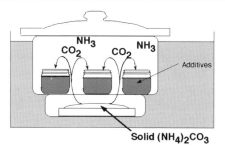

Fig. 3 Experimental setup of a diffusion reactor

in a closed desiccator containing crushed ammonium carbonate. Carbon dioxide is introduced to the solution via vapor diffusion. The crystallization takes place by the slow diffusion of CO_2 into the $CaCl_2$ bath. The carbonate diffusion method was usually applied for crystallization on surfaces of substrates such as self-assembled monolayers of alkylthiols on gold [31–33]. The whole apparatus was placed in an oven or water bath to control the temperature.

Precipitation of calcium carbonate through electrochemistry was proposed to obtain stable and compact covering layers on metal surfaces [34, 35]. The metal surface becomes basic under the influence of the reduction of dissolved oxygen. Consequently, an enrichment of CO_3^{2+} ions, occurs when the local pH increases. The electrochemical environment provides very controlled $CaCO_3$ deposition conditions that are used to govern deposit thickness and deposition rate.

2.2
Amorphous Calcium Carbonate (ACC)

ACC does not persist in a test tube without any additives. ACC rapidly transforms in the presence of water to crystalline polymorphs (Fig. 4). Living organisms, however, can stabilize ACC assisted with specialized matrix molecules and regulate its slow transition to calcite or aragonite. Brečević and Nielsen isolated ACC and found the powder to be non-crystalline and consisting of spherical particles with diameters in the order of 50–400 nm [36]. Aizenberg et al. showed that antler spicule from the branchial sac contain only ACC [37]. Dogbone spicule from the tunic have a core of ACC, surrounded by a sheath [38]. Macromolecules extracted from the calcite layer sped up the formation of calcite crystals from a supersaturated solution. On the other hand, macromolecules extracted from the ACC layer inhibited crystal formation. The addition of macromolecules from the antler spicules favored formation of stable ACC. This constitutes strong evidence that macromolecules occluded with the mineral phase play an important role in stabilization of

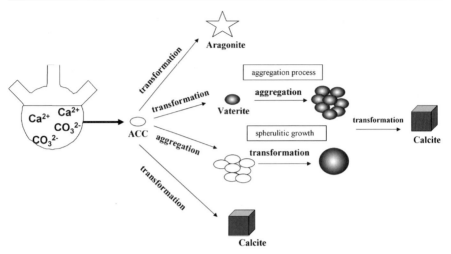

Fig. 4 Schematic depiction for the precipitation of $CaCO_3$ and the transformations from ACC to crystalline phases

relatively unstable forms, like ACC and can also play a role in selection of the polymorph that is formed. The role of ACC in biomineralization has recently been thoughtfully reviewed by Addadi et al. [39]. Organisms use these amorphous phases as building materials, stabilizing them over their life-time, or depositing them as transient phases that transform in a controlled manner into a specific crystalline phase.

2.3
Vaterite

According to the literature, it is difficult to obtain pure vaterite, especially when the ion activity product of the initial supersaturated solution is higher than the solubility product of ACC [40]. Relatively high supersaturations in high pH values favor the precipitation of vaterite [41]. Vaterite is thermodynamically the most unstable of the three crystal structures. It is well known that vaterite transforms into calcite—thermodynamically the most stable—via a solvent-mediated process (Fig. 4) [42]. Even less stable vaterite is present in spicules of some species [43].

Vaterite is expected to be used for various purposes, because it has some interesting features such as high specific surface area, high solubility, high dispersion, and small specific gravity compared with the other two crystal systems. Stable spherical vaterite particles were reported in the presence of divalent cations [44], a surfactant [bis(2-ethylhexyl)sodium sulfate (AOT)] [42], poly(styrenesulfonate) [45], poly(vinylalcohol) [46], double-hydrophilic block copolymers [29], and dendrimers [47–49].

Vaterite nucleation would be kinetically favored over calcite by increases in the rate of cation dehydration at the solution interface with the incipient nuclei [46]. Vaterite particles are usually obtained as a spherically shaped poly crystal, which is built up of 25–35 nm nano-crystallites (aggregation process, see Fig. 4) [44]. Thin slices of the particles were prepared by ultra-microtome for TEM analysis; the size of the individual crystallites should be around 10 nm [50]. The polycrystalline nature of the vaterite particles was also evident by XRD line broadening [44]. In the double-jet precipitation condition, spherical secondary particles of calcium carbonate were formed by controlled agglomeration of primary particles [51]. Much of these works concerned the production of a mono-disperse population in the size-range ~ 0.1–$10\,\mu$m using the double-jet method or by precipitation from a homogenous solution. Matijević and co-workers also suggested the aggregation process mechanism as being responsible for the formation of perfectly spherical particles precipitated directly from solution [52]. Cölfen and Antonietti investigated the precipitation of spherical calcium carbonate microparticles using double-hydrophilic block copolymers and discussed the role of the block copolymers in the aggregation mechanism of nano-crystallites to defined macro-crystals [29].

The spherical shape of vaterite led to other explanations for the formation mechanism. In contrast to the nano-aggregation explanation, spherical crystalline particles were reported to grow by spherulitic growth and by non-epitaxial surface nucleation (Fig. 4) [53, 54]. In these conceptions, spherical particles originate from one single nucleation event and grow from that center by deposition of molecules or ions from the solution. The formation mechanism of vaterite particles in the nucleating environment of spontaneous precipitation experiments was also studied [50]. The experiments were performed by simultaneously mixing equimolar solutions of calcium nitrate and sodium carbonate. Vaterite spheres grew in seeded batch experiments at 25 °C at a moderate initial relative supersaturation in the absence of nucleation. The spherical shape is preserved and the polycrystalline features can be observed on the surface of the particles. The interior of cracked vaterite particles displayed the characteristic radiating feature found in crystals grown by spherulitic growth [53]. Spherulitic growth of calcium carbonate—both vaterite and calcite—has been reported previously, by slow crystal growth in gelatin matrices [55], and reaction between gaseous carbon dioxide from air and calcium in solution [56]. Fluorapatite grown in a gelatin matrix also grows as spherulites [57]. The initial formation of elongated prismatic seeds proceeds through dumbbell shapes to spheres. These dumbbell and peanut shapes are also observed in the precipitation of calcium carbonate, which may indicate that these spherulites of calcium carbonate are initiated by the branching of an elongated prismatic nucleus and not by an isotropic nucleus formed by aggregation of smaller crystals [56]. Dupont reported a method to precipitate mono-crystalline hexagonal platelets of vaterite at 95 °C by suppressing the growth of aragonite and calcite by adding small amounts of surfactant hydroxyethylidene-1,1-phosphonic acid [58].

2.4
Aragonite

Aragonite is metastable under ambient conditions, yet it is found as a natural component. It is formed under a much narrower range of physico-chemical conditions and is easily transformed into calcite by changes in the environment. Specific biological macromolecules are involved in controlling aragonite nucleation [16, 59]. The laboratory synthesis of aragonite from supersaturated solutions at room temperature has hardly been achieved without the use of soluble additives such as Mg^{2+} or small organic molecules [60]. Litvin et al. demonstrated that the spreading of 5-hexadecyloxyisophthalic acid at the air–water interface resulted in the specific nucleation of aragonite from supersaturated calcium bicarbonate solution not doped with additives [61]. They proposed that the geometric match of the template and Ca – Ca distances and angles in the ac plane of aragonite is important. While the slow crystallization of calcium carbonate at 22 °C on self-assembled ω-substituted alkylthiol surfaces yielded the vaterite and calcite, elevating the temperature to 45 °C allowed the stabilization of aragonite [31]. Additives such as Mg^{2+} favor aragonite over calcite by selective kinetic inhibition of the thermodynamically stable calcite structure [62]. The precipitation of aragonite can be achieved by aging solutions of calcium salts in the presence of urea at 90 °C [63]. The typical morphology of aragonite is needle-like crystals with high aspect ratios which are used as fillers for the improvement of mechanical properties of paper and polymer materials [64]. Aragonite is also a suitable biomedical material, because it is denser than calcite and can be integrated, resolved, and replaced by bone [65, 66].

2.5
Calcite

Among the three polymorphs of crystalline calcium carbonate, calcite is the equilibrium phase, most commonly adopting a rhombohedral morphology. In biologically produced minerals, however, the rhombohedral morphology is rarely adopted. Single-crystalline calcite fibers are observed most prominently in sea-urchin teeth and bacterial deposits [67–69]. They do not follow the crystallographic symmetry of calcite. Such calcite fibers were prepared in vitro via a solution-precursor-solid mechanism [70].

3
Multi-Step Nucleation and Growth in the Mineralization Process

3.1
Natural Organism

It is commonly accepted that the protein matrix plays a very important role in regulating biomineral formation. Morphological control can also be accomplished by adsorption of soluble additives onto specific faces of growing crystals, altering the relative growth rates of the different crystallographic faces and leading to different crystal habits. These processes take place usually at an organic–inorganic interface, the organic portion providing the initial structural information for the inorganic part to nucleate on and grow outwards in the desired manner. However, there remain many unknowns as to how the matrix affects the crystallization process, especially the initial nucleation. In a test tube, a crystal precipitates via epitaxy directly from the liquid solution. This concept leads to the belief that matrix controls the crystallization from the very beginning. An initial crystalline phase bears the same structure to a final crystal. Alternatively, the final crystalline phase can arise through a series of steps, initiated by the formation of an amorphous phase that undergoes subsequent phase transformations. Dissection of the crystallization process into several stages could make the activation energy of each step lower than that of the one-step precipitation (Fig. 5). The multistep crystallization process is plausible, especially in a biological environment in which temporal modifications of the crystallization kinetics may be prevalent. Cölfen and Mann proposed the concept of mesoscale transformations and matrix-mediated nucleation in biomineralization [71].

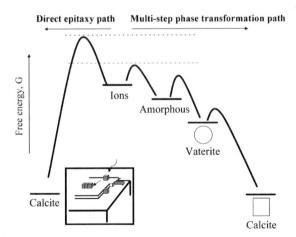

Fig. 5 Energy diagrams of the direct epitaxy path and the multi-step phase transformation path

Important experimental evidence for the multi-step crystallization process on living systems should include detection of different mineral polymorphs in natural organisms and subsequent phase transformation. Weiss et al. showed that both aragonite and ACC were found in the larval shells of two marine bivalves [72]. In *Mercenaria mercenaria*, the initially deposited material is mostly ACC, and this is gradually transformed into aragonite over a period of hours to days. Another molluscan species, *Crassostrea gigas*, also has a mixture of ACC and aragonite in a larval shell, but the shell of the adult is almost all calcite. Hasse et al. examined the mineral in the shell of a developing fresh water snail [73]. The shell of the adults was composed of aragonite only. The mineral in 72-h-old embryos was ACC. ACC apparently transformed to aragonite, which was the first crystalline phase present at 120 h of development. Transformations from the amorphous to the crystalline phase have been discovered and characterized in several cases of biological mineralization [74, 75]. These studies show that the ACC is a precursor to the aragonite, and that the ACC might also be a precursor in the formation of shell in post-metamorphic animals.

As described above, living organisms, presumably employing specialized matrix molecules, can stabilize ACC and regulate its slow transition to calcite or aragonite. Such stabilized ACC may be a precursor for transformation into crystalline forms in biomineralization. In shells with a prismatic layer of calcite and nacre of aragonite, there is a precise control of the $CaCO_3$ polymorph. It is believed that this is due to the matrix molecules that are associated with the forming mineral. Several groups extracted aspartic acid-rich glycoproteins from an aragonitic mollusk shell layer or a calcitic layer [33, 59]. These were aragonite if the added macromolecules were from the aragonitic shell layer, or calcite if they were derived from a calcitic shell layer. In the absence of the acidic glycoproteins, no mineral formed within the complex. This observation leads to the presumption that the cells secreting the matrix are programmed to change the composition of the matrix at precise times and places, thereby regulating the change from calcite to aragonite during the transition from prism to nacre. In nature, stable, crystalline forms of calcium carbonate are formed from an amorphous precursor. This precursor transforms to a crystal in a slow, regulated way. Specific proteins or molecules may be secreted during this transformation to control polymorphs and morphologies of minerals.

3.2
In Vitro Study

Studies seeking evidence for the multi-step in vitro crystallization process have been carried out. Li and Mann were able to demonstrate that in inverse microemulsions, surfactant-vaterite structures with different shapes are formed by a phase transition from stabilized ACC nanoparticles [76]. The addition of water to the ACC nanoparticles caused their self-aggregation and transformation to vaterite. Spontaneous precipitation by the mixing of two concentrated solutions of calcium and carbonate results in ACC as described above. The ACC transformed within 6 min to produce spherical and crystalline vaterite at 25 °C with stirring. The vaterite particles nucleated and grew within the gelatinous ACC [50]. Scanning electron microscopy (SEM) and electrochemical analyses have revealed that the amorphous $CaCO_3$ formed immediately after directly mixing the two solutions of $CaCl_2$ and Na_2CO_3, and subsequently transformed into vaterite and calcite [40, 77].

Much more attention was paid to the transformation process of vaterite to calcite. Upon heating at 730 K, vaterite irreversibly transforms into calcite. This transformation in aqueous solution at ambient conditions was a solution-mediated process [42]. All of the experimental results and the data analysis indicated that the transformation took place through dissolution of vaterite, followed by the crystallization of calcite.

Although aragonite is unstable relative to calcite, aragonite is relatively stable in aqueous solution under moderate conditions. Aragonite transforms into calcite upon heating and on a laboratory time scale this reaction was relatively rapid above 698 K [78]. While solid-state transformation of aragonite to calcite is slow, aragonite was converted to calcite in dilute $CaCl_2$ fluid at temperatures ranging from 50 to 100 °C [79].

3.3
Matrix-Mediated Formation of $CaCO_3$ Thin Films

Groves et al. observed a phase-transformation process from an initially deposited amorphous phase to crystalline calcite during the formation of continuous thin films [80]. They reported the synthesis of macroscopic and continuous carbonate thin films at a porphyrin template/subphase interface by employing poly(acrylic acid) as a soluble inhibitor to mimic the cooperative promotion-inhibition in biogenic thin film production. A semirigid template for crystallization was spontaneously formed via the self-organization of the amphiphilic tricarboxyphenylpophyrin iron(III) μ-oxo dimer at an air–water interface. Calcite crystals that were formed under this porphyrin template were oriented with the (001) face parallel to the template [81]. Films formed at 22 °C were found to have a biphasic structure containing both amorphous and crystalline calcium carbonate. The presence of both the calcite and amorphous calcium carbonate phases in the film is highly suggestive of a phase transformation from the latter to the former. Films obtained in the early stage of formation at lower temperature (4 °C) displayed characteristics of a single amorphous phase. These observations suggested that films formed through a multi-stage assembly process, during which an initial amorphous deposition was followed by a phase transformation into calcite. This study demonstrated that the ACC was a precursor to the calcite films (Fig. 6).

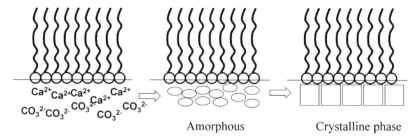

Fig. 6 Schematic of multi-step phase transformation path during formation of $CaCO_3$ films at a template/subphase interface

Zhang et al. employed Langmuir films that formed from amphiphilic molecules with different headgroups at the air–water interface as templates for biomimetic calcium phosphate formation [82]. They observed a phase-transformation process from an initially deposited amorphous phase to a crystalline phase during the initial stage of calcium phosphate formation. If critical nuclei are non-crystalline nuclei, a hydrated crystalline phase is formed first, then a hydrated crystalline phase, and, lastly, an anhydrous crystalline phase. In this process, the nucleation and growth energies are very low. In the initial stage of nucleation of calcium phosphates, non-crystalline nuclei were first formed and grew into big particles. First, Langmuir monolayers with negative charges bound calcium ions, which caused a local concentration of phosphate, which sequentially attracted more calcium ions, which made the concentration of the precursor increase to its supersaturated degree for nucleation. Then, amorphous calcium phosphates were formed. Lastly, phase transformation from amorphous to crystalline calcium phosphates occurred. These observations strongly suggest that the final crystalline phase could arise through a series of steps, initiated by the formation of an amorphous phase that undergoes subsequent phase transformations. Addadi and Weiner et al. have suggested that amorphous calcium phosphate may be much more widespread than the crystalline phase at the beginning of biomineralization process [74, 75].

4
Delayed Action for Nucleation and Growth of $CaCO_3$

The existence of several phases would enable organisms to control mineralization through intervention with kinetics. During the mineralization process in living organisms, specific proteins or molecules may be secreted at a precise time and place to interact with precursor phases. By selectively interacting with the mineral at different stages during the crystal forming process, the organisms could choose to manipulate both the polymorph and the orientation of the mineral to meet specific biological requirements. Although crystallization of calcium carbonate in the presence of the various templates and additives described above has been investigated as a model of biomineralization, selective interaction of organic matrix with the mineral at different stages has been limited. By selectively interacting with the mineral at different stages during the crystal-forming process. The delay addition method provides a new, simple process for controlling the calcium carbonate without synthesizing new additives. In the following sections, examples of the delayed action method are described.

4.1
Overview of Calcium Carbonate Crystallization by Synthetic Substrates

Before the new concept for controlling crystal polymorphs by interaction of organic additives is described, the recent progress of calcium carbonate precipitation by synthetic substrates is briefly discussed. Model systems, in which low molecular-weight organic additives are used to study the effect of molecular properties such as charge and functionality in inorganic crystallization, provide insights into the possible mechanisms operating in biology [46, 83–86]. These additives were chosen to mimic the active protein ligands. Because the proteins that have been found to be associated with biominerals are usually highly acidic macromolecules, simple water-soluble polyelectrolytes, such as the sodium salts of poly(aspartic acid) and poly(glutamic acid), were examined for the model of biomineralization in aqueous solution [87]. Crystallization of $CaCO_3$ in the presence of various synthetic non-peptide polymers has been investigated as a model of biomineralization [17, 18]. For such synthetic linear polymers, it has been difficult to unambiguously assign structure–function relationships in the context of their activity in crystallization assays, since they mostly occur in a random-coil conformation. Dendrimers are monodisperse macromolecules with a regular and highly branched three-dimensional architecture. Because of unique and well-defined secondary structures of the dendrimers, the starburst dendrimers should be a good candidate for studying inorganic crystallization [47–49, 88, 89]. The use of ordered supramolecules, such as micelles [90], monolayers [91], vesicles [92], inverted micelles [93], and lyotropic liquid crystalline systems [94], also allow for controlled mineralization.

4.2
Delay Addition of Poly(acrylic acid)

The influences of the sodium salt of poly(acrylic acid) (PAA) on the crystallization of $CaCO_3$ to act as an inhibitor for crystal formation has been intensively investigated [80, 95, 96]. In the presence of PAA under a feed ration of a repeating unit of acrylate to calcium ions of 0.62 using the double-jet method, little precipitate was collected after incubation at 25 °C for 4 days [97]. The IR spectrum of the $CaCO_3$ obtained with less than 1% yield in the presence of PAA showed an amorphous character. The aggregation of PAA and $CaCO_3$ particles formed in solution was proposed [98]. The samples were taken one hour after the start of the mineralization process by allowing carbon dioxide to diffuse into the solution. The sample contained small $CaCO_3$ particles with a size of about 10 nm by HR-TEM measurements. Samples taken about 20 min after the start of the mineralization process seemed to be amorphous.

A delay addition of sodium salt of PAA for precipitation of $CaCO_3$ was carried out by the double-jet method [97]. After addition of the aqueous solutions of 0.1 M $CaCl_2$ and 0.1 M $(NH_4)_2CO_3$ into an aqueous solution, an aqueous solution of sodium salt of PAA (M_w = 1200) was added to the reaction mixture after incubation at 30 °C for several minutes. The feed ratio of a repeating unit of PAA to calcium ions was 0.62. Little crystalline $CaCO_3$ was collected after incubation at 30 °C under N_2 for 1 day when the sodium salt of PAA (M_w = 1200) was added to the reaction mixture after incubation at 30 °C for 1 min. When the sodium salt of PAA (M_w = 1200) was added to the reaction mixture after incubation for 20 min, a small amount of vaterite was obtained.

High pH promotes the rate of crystallization of calcium carbonate. Before adding the calcium reactants, the water in a precipitation flask was adjusted to pH 8.5 with NH_4OH. In the initial presence of the sodium salt of PAA (M_w = 5000), no precipitate was collected under a feed ration of a repeating unit of acrylate to calcium ions of 0.62 using the double-jet method. After both the aqueous solutions of 0.1 M $CaCl_2$ and 0.1 M $(NH_4)_2CO_3$ were injected into the water, an aqueous solution of sodium salt of PAA was added to the reaction mixture after incubation for several minutes. Although PAA initially acted as an inhibitor for nucleation and growth of crystallization, stable vaterite particles were obtained by the delaying addition of PAA after 1 to 60 min (Haung et al., 2006, personal communication). The resulting spherical vaterite particles were stable in an aqueous solution for more than 1 week. These results indicate that the final crystalline phases are highly sensitive to the presence of the active additives at the very initial nucleation stage.

4.3
In-situ Polymerization

Another new concept for controlling a polymorph of calcium carbonate as schematically shown in Fig. 7 was proposed by Naka et al. [99]. The key point of the method is using a "latent inductor" for crystal nucleation. The latent inductor at the inactive state does not affect the nucleation and growth of a crystal. After the inactive state is transferred to an active state by a stimulus, the active inductor can control the nucleation and growth of the crystal. Sodium acrylate was used as a latent inductor for this purpose and potassium

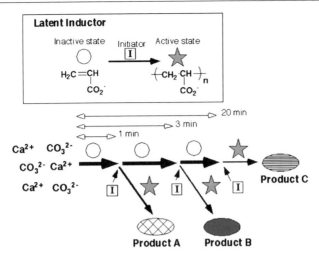

Fig. 7 Schematic depiction for the control of crystal polymorph by a latent inductor

peroxodisulfate was used as a stimulus. Sodium acrylate may not affect the nucleation and growth of the crystal [100]. On the other hand, poly(acrylate) acts as an inhibitor for crystal formation as described in Sect. 4.2. Sodium acrylate can be transferred to poly(acrylate) by adding the radical initiator. It should be noted that crystallization of calcium carbonate with in-situ polymerization of anionic monomers in an aqueous solution has not been reported so far.

The precipitation of $CaCO_3$ was carried out using the double-jet method. After the addition of the calcium reactants into the aqueous solution of sodium acrylate was completed, an aqueous solution of potassium peroxodisulfate (KPS) as a water-soluble radical initiator was added to the reaction mixture after incubation at 30 °C for several minutes (1, 3, or 20 min). This solution was then kept at 30 °C and the critical point of the sudden increase in the turbidity of the solution was observed at 3 to 4 min of stirring after the addition of the calcium solution. When the radical initiator was added to the calcium solution with sodium acrylate after incubation for 1 min, no turbidity of the solution was observed. These solutions were kept at 30 °C under N_2 for 1 day with gentle stirring. Three different crystal polymorphs of $CaCO_3$ (aragonite, vaterite, and calcite) were selectively induced by changing the time at which the radical initiator was added to a calcium carbonate solution with sodium acrylate (Table 1). Figure 8 shows the scanning electron micrographs (SEM) of the three crystalline products. Most crystals of run 1 of Table 1 were efflorescent bundles of needles (Fig. 8a), which is a typical aragonite crystal morphology. The product of run 2 consisted of two different crystal modifications: spherical vaterite and rhombs of calcite (Fig. 8b). The crystal of the product of run 3 was rhombohedral (Fig. 8c). Each shape of $CaCO_3$ is a typi-

Table 1 Formation of crystalline $CaCO_3$ in the presence of sodium acrylate with a radical initiator at different feed ratios of sodium acrylate to calcium ions at 30 °C for 1 day. (Adapted from [97])

Run	[acrylate]/[Ca^{2+}]	Addition time of KPS/min	Yield/%	Polymorphism[a]
1	0.62	1	33	aragonite + calcite (trace)
2	0.62	3	35	vaterite (63%) + calcite
3	0.62	20	54	calcite

[a] Polymorphism was characterized by FT-IR. The fraction of vaterite in the crystalline phase was determined by XRD

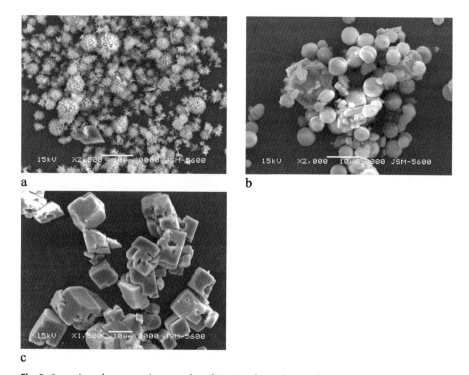

Fig. 8 Scanning electron micrographs of $CaCO_3$ from the products of run 1 (**a**), run 2 (**b**), and run 3 (**c**) in Table 1. (Reproduced with permission from [97]. Copyright 2004 The Chemical Society of Japan)

cal morphology for each polymorph. The thermogravimetric analysis showed that organic additives were not built into the crystals obtained to any significant degree after washing with water.

The crystalline product obtained with sodium acrylate without addition of the radical initiator was calcite. The mono-carboxylic acid did not exert any influence on the nucleation and crystal growth of $CaCO_3$. The crystal phase of $CaCO_3$ obtained without any additives was also calcite. Sodium acrylate was regarded as an inactive form for induction of metastable $CaCO_3$ crystalline phases (vaterite or aragonite).

The crystal polymorph of the product of run 2 did not change when the solution was kept for 2 days. Since the vaterite crystal was transformed to calcite when the solution was incubated for 3 days, this suggested that the vaterite surfaces were stabilized by the resulting poly(acrylate) in aqueous solution to prevent phase transformation (Fig. 9). The formation of aragonite is usually achieved at a higher temperature than 50 °C using a solution method of preparation. In these results, aragonite can be formed at 30 °C when the radical initiator was added to the calcium solution with sodium acrylate after incubation for 1 min, in which the $CaCO_3$ crystal formation had not started. It is possible that aragonite is rapidly nucleated at the very beginning of the nucleation process, resulting in being kinetically induced by poly(acrylate) (Fig. 9). The present result provides a new method for aragonite formation at ambient temperature in a homogeneous nucleation system. In addition, the crystal phase of the product of run 3 in Table 1 was also thermodynamically stable calcite. The final crystalline phase was not affected by the polymerization of sodium acrylate after incubation for 20 min.

A higher temperature of 35 °C was employed for crystallization of calcium carbonate with radical polymerization of sodium acrylate in aqueous solution. When the concentration of sodium acrylate was the same as that of Table 1, the polymorphs of the products obtained when the radical initiators were added to the reaction mixture after incubation for 1 min (run 1) and 3 min (run 2) were amorphous and aragonite, respectively (Table 2). Aragonite formation was also observed when the radical initiators were added to the reaction mixture after incubation for 20 min. These results suggest that

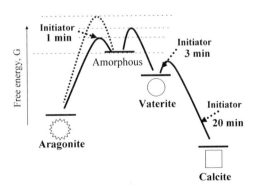

Fig. 9 Crystallization pathways under the in-situ polymerization of acrylate

Table 2 Formation of crystalline $CaCO_3$ in the presence of sodium acrylate with a radical initiator at different feed ratios of sodium acrylate to calcium ions at 35 °C for 1 day. (Adapted from [97])

Run	[acrylate]/[Ca^{2+}]	Addition time of KPS/min	Yield/%	Polymorphism [a]
1	0.62	1	< 1	amorphous
2	0.62	3	29	aragonite
3	0.62	20	21	aragonite

[a] Polymorphism was characterized by FT-IR

higher temperature tends to induce aragonite formation due to reduced activation energy for aragonite nucleation.

Poly(N-vinylpyrrolidone) (PVP) is a water-soluble and uncharged polymer. The presence of PVP has no influence on the polymorphs of $CaCO_3$ precipitation, but has a morphological effect on vaterite and calcite at high PVP concentration [101]. The precipitate obtained in the initial presence of PVA was calcite (run 2 of Table 3). The crystalline products obtained with N-vinylpyrrolidone without addition of the radical initiator were calcite with a trace amount of aragonite (run 1 of Table 3). These results indicate that both the polymer and monomer did not exert any influence on the nucleation and crystal growth of $CaCO_3$. On the contrary, in-situ polymerization of the monomer during the precipitation of $CaCO_3$ was carried out by the double-jet method (Keum et al., 2006, personal communication). After addition of the calcium reactants into the aqueous solution of the monomer was completed, an aqueous solution of KPS as a water-soluble radical initiator was added to the reaction mixture after incubation at 30 °C for several minutes (1, 3, or 20 min). All the products obtained by the in-situ polymerization were pre-

Table 3 Formation of crystalline $CaCO_3$ with in-situ radical polymerization of N-vinyl-2-pyrrolidone at 30 °C

Run	[−COONa]/[Ca^{2+}]	Addition time (min)	Polymorphism [a]
1	0.75	monomer	Calcite ≫ Aragonite
2	0.75	polymer	Calcite
3	0.75	1	Aragonite ≫ Calcite
4	0.75	3	Aragonite ≫ Calcite
5	0.75	20	Aragonite ≫ Calcite

[a] Polymorphism was characterized by FT-IR

dominantly aragonite. On the basis of the results that both the initial presence of the monomer and polymer has no specific influence on the polymorphs of $CaCO_3$, the final crystalline phases are highly affected by the in-situ polymerization at the very initial nucleation stage. It is possible that aragonite is rapidly nucleated at the very beginning of the nucleation process, resulting in being kinetically induced by the polymerization.

4.4
Delay Addition of Anionic Polyamidoamine Dendrimers

The anionic polyamidoamine (PAMAM) dendrimers act as effective protective agents for the most unstable vaterite crystal [47–49]. Delay addition of the anionic PAMAM dendrimers also gave vaterite particles by the double-jet method [102]. After the calcium reactants (0.1 M $CaCl_2$ and 0.1 M $(NH_4)_2CO_3$) were injected via syringe into distilled water, an aqueous solution of the G1.5 PAMAM dendrimer was added to the reaction mixture after the start of precipitation for several minutes (1, 3, 20, or 60 min). The experimental conditions and results are summarized in Table 4. At the high

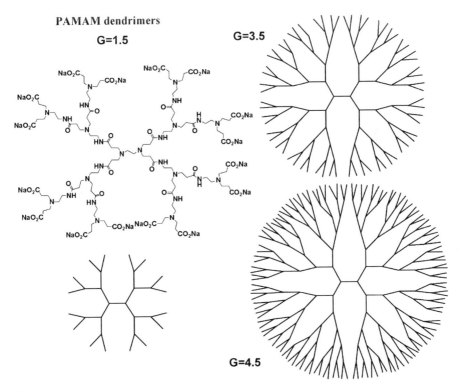

Structure 1

Table 4 Formation of crystalline CaCO$_3$ with the G1.5 PAMAM dendrimer at different addition times and different feed ratios of the dendrimer to calcium ions at 30 °C. (Adapted from [102])

Run	[−COONa] (mM)	[−COONa]/[Ca^{2+}]	Addition time (min)	Polymorphism [a]	Particle Size of Vaterite [b] (μm)	Yield (%)
1	0.27	0.1	0	Vaterite	2.3±1.1	23
2	"	"	1	Vaterite (93%) [c] + Calcite	2.8±1.2	36
3	"	"	3	Vaterite (90%) [c] + Calcite	2.1±0.6	28
4	"	"	20	Vaterite + Calcite + Aragonite	9.1±0.2	27
5	"	"	60	"	6.7±0.7	31
6	1.37	0.5	0	Vaterite	2.2±1.6	26
7	"	"	1	"	1.7±0.6	44
8	"	"	3	Vaterite (92%) [c] + Calcite	2.4±1.8	40
9	"	"	20	Vaterite (87%) [c] + Calcite	4.6±0.6	25
10	"	"	60	Vaterite + Calcite + Aragonite	6.9±0.4	29
11	2.75	1	0	Vaterite	1.2±0.5	22
12	"	"	1	"	1.3±0.3	48
13	"	"	3	"	3.2±0.9	44
14	"	"	20	"	3.8±0.2	20
15	"	"	60	"	4.7±0.7	47

[a] Polymorphism was characterized by FT-IR
[b] Particle size was measured by SEM
[c] Vaterite content was determined by XRD

concentration of the G1.5 PAMAM dendrimer corresponding to 2.75 mM of −COONa, all five samples were vaterite. As the concentration of −COONa decreased to 1.37 mM, the obtained crystal phases were vaterite when the dendrimer was added after incubation for 0 and 1 min. However, calcite coexisted with vaterite when the dendrimer was added after incubation for 3 min. Although vaterite was predominantly formed even at the lower concentration of the G1.5 PAMAM dendrimer corresponding to 0.27 mM of −COONa, the amounts of calcite and aragonite forms were increased with an increase in the delay addition time of the dendrimer. On the basis of the fact that the obtained crystal phase in the absence of the dendrimer was calcite, the dendrimers effectively stabilized the most unstable vaterite crystal even delaying the addition time of the dendrimer for 60 min. When the dendrimer was added after incubation for 120 min, calcite was predominantly formed.

The most interesting characteristic of the delay addition is the size of the vaterite particles. As the delay addition time of the G1.5 PAMAM dendrimer increased from 0 to 60 min, the particle size of the spherical vaterite increased from 1.2 ± 0.5 to 4.7 ± 0.7 µm (Fig. 10). Furthermore, as the concentration of –COONa decreased, the particle sizes of the spherical vaterite increased. The mixtures of calcite and aragonite were also increased when the lower concentration of the PAMAM dendrimer was used under the same conditions. These results indicate that the inhibitor effect of the dendrimer was weakened by decreasing the concentration of the PAMAM dendrimer.

The precipitations of $CaCO_3$ in the presence of the G3.5 PAMAM dendrimer were also carried out under the same conditions. The results are summarized in Table 5. No calcite formation was observed in all the concentrations. As the generation number of the anionic PAMAM dendrimer increased from G1.5 to G3.5, the particle sizes of the spherical vaterite de-

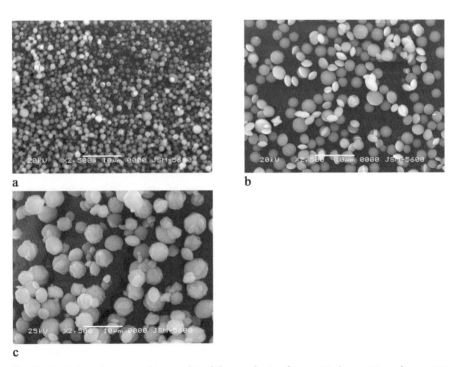

Fig. 10 Scanning electron micrographs of the products of **a** run 11, **b** run 13, and **c** run 15 in Table 4. (Reproduced with permission from [102]. Copyright 2003 The Chemical Society of Japan)

Table 5 Formation of crystalline $CaCO_3$ with G3.5 PAMAM dendrimer at different addition times and different feed ratios of the dendrimer to calcium ions at 30 °C. (Adapted from [102])

Run	[−COONa] (mM)	[−COONa]/ [Ca^{2+}]	Addition time (min)	Polymorphism [a]	Particle Size of Vaterite [b] (μm)	Yield (%)
1	0.27	0.1	0	Vaterite	1.3±0.2	27
2	″	″	1	″	1.4±0.8	25
3	″	″	3	″	1.5±0.8	29
4	″	″	20	Vaterite+Aragonite	6.2±0.5	24
5	1.37	0.5	0	Vaterite	1.2±0.7	24
6	″	″	1	″	1.4±0.4	36
7	″	″	3	″	1.6±0.3	28
8	″	″	20	″	5.5±0.8	23
9	2.75	1	0	Vaterite	1.1±0.6	17
10	″	″	1	″	1.1±0.3	18
11	″	″	3	″	1.2±0.5	23
12	″	″	20	″	3.1±1.7	21

[a] Polymorphism was characterized by FT-IR
[b] Particle size was measured by SEM

creased under the same conditions. These results suggest that the G3.5 dendrimer effectively stabilized the vaterite particles compared with the earlier generation of the dendrimer. The surface of the nanosized metastable vaterite might be modified with the PAMAM dendrimer. The later generation of the dendrimer effectively protected the vaterite particles compared with the earlier generation [48].

The precipitation of $CaCO_3$ in the absence of any additives under comparable condition was continued. Crystalline products were immediately isolated after incubation for 3 and 20 min by centrifugation. In the case of the product after an incubation period of 3 min, vaterite and calcite coexisted. The size of the vaterite particle was 4.2 ± 1.6 μm (Fig. 11a). In the case of the product after 20 min, the size of the vaterite particle was 10 ± 2.3 μm (Fig. 11b). The size of the spherical vaterite increased with an increase in incubation time. These results suggest that the anionic PAMAM dendrimers effectively modified the vaterite surface and inhibited further growth of the vaterite particles. The higher concentration of the PAMAM dendrimer adsorbed on the vaterite particles of $CaCO_3$ more completely and smaller-sized vaterite was produced compared with that in the lower concentration of the PAMAM dendrimer. These results suggest that the PAMAM dendrimers act as effective stabilizers for the metastable vaterite phase, and

Fig. 11 Scanning electron micrographs of CaCO$_3$ isolated after incubation for 3 min (**a**) and 20 min (**b**) in the absence of any additives. (Reproduced with permission from [102]. Copyright 2003 The Chemical Society of Japan)

quenched the aggregation of vaterite crystals by changing the addition time of the dendrimer.

4.5
Polymers Responsive to Stimulation

Functional polymers that possess responsive properties to a certain stimulation have been well developed. Stimulation includes physical and chemical stimulation such as light, heat, electricity, magnetic field, pH, ions, and bioactive molecules. In the in-situ radical polymerization of acrylate described in Sect. 4.3, the radical initiator is regarded as a chemical stimulation for transformation from an inactive additive to an active form. After the inactive state is transferred to an active state by a stimulus, the active inductor can control the nucleation and growth of the crystal. This concept can apply to other responsive materials as latent inductors. Materials with reversible conversion of properties by light are called photochromic materials. Photochromism is defined as a reversible transformation of a single chemical species between two states having different distinguishable absorption spectra.

Naka et al. reported on the control of vaterite shapes by photo-induced *cis-trans* isomerization of an azobenzene-containing polymer in a mixture of dimethyl sulfoxide (DMSO) and water [103]. Photoisomerization of azobenzene derivatives is well known as a configurational switch in photoreactions [104]. In addition, the *trans* isomer shows an intense absorption around 320 nm by the π–π^* transition, and the *cis* isomer has a weak absorption of the n–π^* transition around 430 nm. This photoisomerization of the azobenzene groups can be used to trigger a change of properties such as polarity, free volume, and conformation [105–108]. Many studies of polymers containing azobenzene groups have been applied for a variety of smart materials [109, 110].

Structure 2

Crystallization of $CaCO_3$ was performed by adding $(NH_4)_2CO_3$ via a syringe into a mixture of water and DMSO solution (20%) containing poly[1-[4-(3-carboxy-4-hydroxyphenylazo)benzenesulfonamido]-1,2-ethanediyl sodium salt] (PAZO) and $CaCl_2$ at 30 °C. After this addition, a sudden increase in the turbidity of the solution was observed after incubation for 3 min. The experimental conditions and the results are summarized in Table 6. In the

Table 6 Shape control of spherical vaterite particles by photo-induced *cis–trans* isomerization of the azobenzene-containing polymer in a mixture of DMSO and water. (Adapted from [103])

Run	Additive	Weight of additive [mg]	Polymorphism [a]	Yield [%] [b]
1	PAZO (Trans)	4.0	Vaterite (68%) + Calcite	41.2
2	" (T → C)	"	Vaterite (84%) + Calcite	35.5
3	" (Cis)	"	Vaterite (77%) + Calcite	38.8
4	AYG [c] (Trans)	3.8	Vaterite (73%) + Calcite	43.6
5	.	.	Vaterite (72%) + Calcite	45.9

[a] Polymorphism was characterized by XRD
[b] Estimated on the basis that all the calcium carbonate was recovered
[c] Alizarin Yellow GG was used instead of PAZO

case of run 2, the reaction mixture was irradiated by a UV light after incubation for 1 min. The product of run 3 was obtained by irradiation by UV light for 1 h before addition of $(NH_4)_2CO_3$ into the solution and the UV light was continuously irradiated to the solution during crystallization of $CaCO_3$ for 3 h.

In the presence of the *trans* isomer of PAZO, the crystals of run 1 were aggregates of disk-shaped vaterite (Fig. 12a). The spherical vaterite was changed to irregular vaterite having a flatter surface with increased size than that of run 1 by the in-situ photoisomerization of PAZO after incubation for 1 min (Fig. 12c). In the case of run 3, the mixture of oval-shaped vaterite with irregular vaterite (about 30%) was obtained in the presence of the *cis* isomer of PAZO (Fig. 12e). Polycrystalline features can be observed on the surface of the oval-shaped vaterite. The *trans* isomer of Alizarin Yellow GG (AYG) was used as a unit model instead of PAZO under the same conditions (run 4), the product was a mixture of center hole vaterite spheres with spherical vaterite with a rough surface (Fig. 12g). In the absence of additives (run 5), in the product of $CaCO_3$ the vaterite crystals were not so different in shape to those of run 3 and 4 (Fig. 12h).

Generally, vaterite formation is easily produced in non-aqueous solvents from the acceleration of spontaneous precipitation rate with the stabilizer effect for protection of vaterite transformation to calcite. The spherical vaterite of rough surface was produced in the presence of the *trans* isomer in PAZO without UV irradiation (run 1). Interestingly, the shape of the crystals from run 1 was changed to irregular vaterite with a flatter surface by the in-situ irradiation of the UV light after incubation for 1 min under the same conditions as shown above (run 2). The product of run 1 was produced by aggregation of limited small particles from the nucleation on the *trans* isomer in PAZO. In the case of run 2, however, nucleation of the primary particles might be inhibited by configurational change of PAZO from the *trans* isomer to the *cis* isomer via irradiation by UV light after incubation for 1 min (Fig. 13). The rearrangement of unstable amorphous $CaCO_3$ colloids at the initial stage simultaneously occurred with isomerization of PAZO at the interface. Thus, the unstable amorphous colloids of primary particles diffused along the edges for a second growth of $CaCO_3$ crystals by the inhibitor effect of PAZO. In addition, the increased size of run 2 (12.2 ± 0.3 μm) shows evidence of a more retarded nucleation rate than that of run 1 (9.5 ± 0.7 μm). It is well known that fast nucleation compared to growth produces small particle sizes in colloid chemistry. The product of run 3 was oval-shaped vaterite with about 30% irregular vaterite similar to run 2. The 30% irregular vaterite should be produced by generally reversible *cis–trans* isomerization of the azobenzene group from the thermodynamically unstable *cis* isomer to the stable *trans* isomer. However, the size and shape of the crystals in the remaining 70% of vaterite are similar to those of run 5. The result indicates that the inhibitor effect of the anionic group of the *cis* isomer for the $CaCO_3$ crystallization might

Fig. 12 Scanning electron micrographs of the products of **a** run 1, **c** run 2, **e** run 3, **g** run 4, and **h** run 5 with higher magnification images of the products of **b** run 1, **d** run 2, and **f** run 3 in Table 6. (Reproduced with permission from [103]. Copyright 2004 Elsevier)

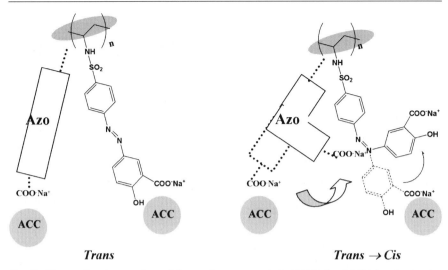

Fig. 13 The molecular motion of the azobenzene unit in PAZO for inhibition by in situ photo-induced isomerization. (Reproduced with permission from [103]. Copyright 2004 Elsevier)

be lower than that of the *trans* isomer, due to the interior position of the azobenzene group of the *cis* isomer as shown in Fig. 13. These results indicated that the reversible isomerization in molecular levels of the azobenzene groups induced configurational change of the template for modified shapes of vaterite.

5
Template Mineralization on the Surface of Calcium Carbonate

The anionic PAMAM dendrimers, which are spherical and proposed as mimics of anionic proteins, adsorbed strongly on the surfaces of metastable vaterite particles [47, 48]. The dendrimers are regarded as "spherical linkers" for surface functionalization of $CaCO_3$. The spherical linkers are rigid and completely defined, and functional groups are unable to attach to the same inorganic surfaces due to the steric hindrance of the spherical structures. The strong bonding to the inorganic surfaces is also expected due to a chelate or cluster effect. The spherical linkers on the surface of minerals provided interfacial active sites for nucleation of the second crystalline phase to produce a hierarchical crystal growth as shown in Fig. 14. Anionic-functionalized metal nanoparticles are also candidates for such a purpose.

Naka et al. used tiopronin-protected gold nanoparticles (Au-Tiopronin) as spherical linkers to prepare surface-functionalized particles and provided for the template mineralization of $CaCO_3$ to construct sea urchin-shaped struc-

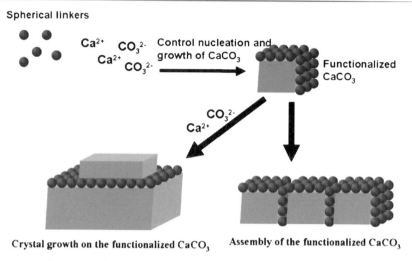

Fig. 14 Template mineralization on the surface of functionalized $CaCO_3$

Fig. 15 SEM images of the time-dependent crystal growth of the product by isolation of the crystals after incubation for 10 (**a**), 20 (**b**), and 90 min (**c**), respectively. (Reproduced with permission from [111]. Copyright 2004 The Chemical Society of Japan)

tures [111]. Time-dependent crystal growth of products in the presence the anionic gold nanoparticles was monitored under the double-jet precipitation method. The crystals isolated after 10 min incubation (Fig. 15a) were brown in color and spherical vaterite was produced from the aggregation of disc vaterite with about 3 µm diameter. These vaterite particles did not transform into thermodynamically more stable forms in contact with water for more than 1 week. The results indicate that these vaterite particles were stabilized by Au-Tiopronin to prevent phase transformation to calcite. After isolation at 20 min incubation, tiny crystals were grown on the surface of the spherical vaterite (Fig. 15b). Finally, the sea urchin-shaped $CaCO_3$ was formed after incubation for 90 min (Fig. 15c). The brown color of the sample faded out in the case of the crystalline products isolated after further incubation.

Au-Tiopronin on the surface of the vaterite particles provided interfacial active sites for nucleation of the second crystalline phase to produce a hierarchical crystal growth with crystal phase switching and the remaining Au-Tiopronin controlled the second crystal growth with selective interaction.

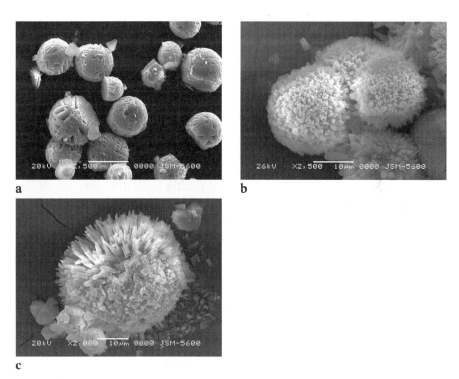

Fig. 16 SEM images of **a** the surface-functionalized vaterite particles and **b,c** the products from template mineralization on the surface-functionalized vaterite particles. The amounts of calcium ions were 0.49 mM (**b**) and 0.78 mM (**c**), respectively. (Reproduced with permission from [111]. Copyright 2004 The Chemical Society of Japan)

The presence of Au-Tiopronin also played an important role for promoting crystal growth of the needle-shaped aragonite on the surface of the functionalized vaterite particles. The presence of Au-Tiopronin in the solution induced the second nucleation and growth of $CaCO_3$ crystals on the surface of the gold nanoparticle-functionalized vaterite particles.

Surface-functionalized vaterite particles with Au-Tiopronin were formed by increasing the concentration of Au-Tiopronin compared to the product described above. The template mineralization was carried out in the presence of the surface-functionalized vaterite particles. The calcium reactants were injected via syringe into an aqueous solution in the presence of the surface-functionalized vaterite particles with the additional Au-Tiopronin. The product was isolated after incubation for 1 day. The sea urchin-shaped $CaCO_3$ with a little rhombohedral calcite was observed by SEM (Fig. 16). From the SEM images, needle-shaped crystals on the surface of the vaterite particles were elongated which increasing the amount of the calcium reactants. This process helps with understanding how to develop the new biomimetic materials and with the template mineralization mechanism.

6
Conclusions and Outlook

Over the past decade, a considerable amount of work has been carried out on the mineralization of calcium carbonate in the presence of various organic templates and additives. This interesting research effort has led to fundamental developments in areas relating to the biomineralization process. Natural inorganic–organic hybrid materials are manufactured through very complexed multi-step-processes. First, the extracellular macromolecular substrates are pre-organized for regiospecific nucleation and the subsequent development of biominerals with controlled micro-architecture. Second, inorganic ions are selectively transported to produce localized high-concentrations within discrete organized compartments. Third, crystallization of inorganic materials is selectively induced at the site of biomineralization. The induction of crystallization occurs by specific matrix. Crystal morphology, size, polymorph, and orientation are controlled by local conditions and, in particular, the presence of "matrix" proteins or other macromolecules. Finally, the crystal growth is determined by the pre-organized cellular compartments. Most biomineralization studies were focused on the third stage.

The protein matrix plays a very important role in regulating biomineral formation. Morphological control can also be accomplished by adsorption of soluble additives onto specific faces of growing crystals, altering the relative growth rates of the different crystallographic faces and leading to different crystal habits. The final crystalline phase arises through a series of steps initiated by the formation of an amorphous phase that undergoes subsequent phase transformations. The multi-step crystallization process on living systems was supported by the detection of different mineral polymorphs in natural organisms and subsequent phase transformation. During this transformation, specific proteins or molecules may be secreted at a precise time and place to control polymorphs and morphologies of minerals. By selectively interacting with the minerals at different stages during the crystal forming process, the organisms may choose to manipulate both the polymorph and the orientation of the mineral to meet specific biological requirements.

These observations of nature have inspired us to find a new mineralization method to develop inorganic–organic hybrid materials. During the formation of continuous thin films of minerals, several authors have used a phase-transformation process from an initially deposited amorphous phase to a crystalline phase. The delay addition method gives a new simple process for controlling the $CaCO_3$ crystallization. Three different crystal polymorphs of $CaCO_3$ (aragonite, vaterite, and calcite) were selectively induced by changing the time when the radical initiator was added to a calcium carbonate solution with sodium acrylate. These processes may be similar to the secretion of specific proteins or molecules during the transformation of biomineralization.

Although many exciting and excellent works have been produced on mineralization in the presence of various synthetic additives and matrix, the naturally produced minerals are superior to these artificial products. While there can be no doubt that these crystal structures are determined by local conditions in natural systems, the specific biological requirements of the organisms to manipulate both the polymorph and the orientation of the mineral are open to question. These questions are related to the genetic basis for the diversity and evolution of biominerals. In my opinion, this point may be clarified during ongoing and continuous efforts in the study of the fundamentals of biomineralization. The continuous cooperation of organic and polymer chemists with inorganic and biochemists is desirable to seek new concepts and methods for constructing composite materials and crystalline forms analogous to those produced by nature. I expect that these continuing and fundamental efforts directed at clarifying the biomineralization process in nature will lead to the next industrial revolution.

References

1. Schmid G, Maihack V, Lantermann F, Peschel S (1996) J Chem Soc, Dalton Trans, p 589
2. Beecroft LL, Ober CK (1998) Chem Mater 10:1440
3. Sanchez C, Lebeau B, Chaput F, Boilot JP (2003) Adv Mater 23:1969
4. Usuki A, Kawakami M, Kojima Y, Fukushima Y, Okada A, Kurauchi T, Kamigaito O (1993) J Mater Res 8:1179

5. Jousseaume B, Lahcini M, Rascle MC, Ribot F, Sanchez C (1995) Organometallics 14:685
6. Tamaki R, Tanaka Y, Asuncion MZ, Choi J, Laine RM (2001) J Am Chem Soc 123:12416
7. Kageyama K, Tamazawa J, Aida T (1999) Science 285:2113
8. Naka K, Yaguchi M, Chujo Y (1996) Chem Mater 11:849
9. Chujo Y, Saegusa T (1992) Adv Polym Si 100:11
10. Chujo Y (1996) Polym Mater Encycl 6:4793
11. Chujo Y (1996) Curr Opin Solid State Mater Sci 1:806
12. Smith BL (1999) Nature 399:761
13. Song F, Bai YL (2003) J Mater Res 18:1741
14. Kumar W, Nukala VV, Simunović (2005) Physic Rev E 72:041919
15. Addadi L, Weiner S (1985) Proc Natl Acad Sci USA 82:4110
16. Falini G, Albeck S, Weiner S, Addadi L (1996) Science 271:67
17. Naka K, Chujo Y (2001) Chem Mater 13:3245
18. Estroff LA, Hamilton AD (2001) Chem Mater 13:3227
19. Gowner LA, Tirrell DA (1998) J Cryst Growth 191:153
20. Cölfen H, Qi L (2001) Chem Eur J 7:106
21. Manoli F, Dalas E (2001) J Cryst Growth 222:293
22. Kahmi SR (1962) Acta Cryst 16:770
23. Kazmierczak F, Tomson MB, Nancollas GH (1982) J Phys Chem 86:103
24. House WA (1981) J Chem Soc Faraday Trans 1 77:341
25. Cassford GE, House WA, Pethybridge AD (1983) J Chem Soc Faraday Trans 1 79:1617
26. Kavanagh AM, Rayment T, Proce TJ (1990) J Chem Soc Faraday Trans 1 86:965
27. Clarkson JR, Price TJ, Adams CJ (1992) J Chem Soc Faraday Trans 88:243
28. Stávek J, Sípek M, Hirasawa I, Toyokura K (1992) Chem Mater 4:545
29. Cölfen H, Antonietti M (1998) Langmuir 14:582
30. Sedlák M, Antonietti M, Cölfen H (1998) Macromol Chem Phys 199:247
31. Küther J, Tremel W (1997) Chem Commun, p 2029
32. Küther J, Seshadri R, Knoll W, Tremel W (1998) J Mater Chem 8:641
33. Küther J, Seshadri R, Nelles G, Assenmacher W, Butt HJ, Mader W, Tremel W (1999) Chem Mater 11:1317
34. Simpson LJ (1998) Electrochemica Acta 43:2543
35. Tlili MM, Benamor M, Gabrielli C, Perrot H, Tribollet B (2003) J Electrochemi Sci 150:C765
36. Brečević L, Nielsen AE (1989) J Crystal Growth 98:504
37. Aizenberg J, Lambert G, Addadi L, Weiner S (1996) Adv Mater 8:222
38. Aizenberg J, Lambert G, Weiner S, Addadi L (2002) J Am Chem Soc 124:32
39. Addadi L, Raz S, Weiner S (2003) Adv Mater 15:959
40. Ogino T, Suzuki T, Sawada K (1987) Geochimica et Cosmochimica Acta 51:2757
41. Spanos N, Koutsoukos PG (1998) J Phys Chem B 102:6679
42. Lopezmacipe A, Gomezmorales J, Rodriguezclemente R (1996) J Cryst Growth 166:1015
43. Lowenstam HA, Abbott DP (1975) Science 188:363
44. Brečević L, Nothing-Laslo V, Kralji D, Popovic S (1996) J Chem Soc Faraday Trans 92:1017
45. Kawaguchi H, Hirai H, Sakai K, Nakajima T, Ebisawa Y, Koyama K (1992) Colloid Polym Sci 270:1176
46. Didymus JM, Oliver P, Mann S, Devries AL Hauschka PV, Westbroek P (1993) J Chem Soc Faraday Trans 89:2891
47. Naka K, Tanaka Y, Chujo Y, Ito Y (1999) Chem Commun, p 1931

48. Naka K, Tanaka Y, Chujo Y (2002) Langmuir 18:3655
49. Naka K, Kobayashi A, Chujo Y (2002) Bull Chem Soc Jpn 75:2541
50. Andreassen JP (2005) J Cryst Growth 274:256
51. Stávek J, Sípek M, Hirasawa I, Toyokura K (1992) Chem Mater 4:545
52. Privmann V, Goia DV, Park J, Matijevic E (1999) J Colloid Int Sci 213:36
53. Goldenfeld N (1987) J Crystal Growth 84:601
54. Sugimoto T, Dirige GE, Muramatsu A (1996) J Colloid Int Sci 182:444
55. Grassmann O, Löbmann P (2004) Biomaterials 25:277
56. Meldrum FC, Hyde ST (2001) J Crystal Growth 231:544
57. Busch S (1999) Eur J Inorg Chem 1643
58. Dupont L, Portemer F, Figlarz M (1997) J Mater Chem 7:797
59. Belcher AM, Wu XH, Christemsen RJ, Hansma PK, Stucky GD, Morse DE (1996) Nature 381:56
60. Heywood BR, Mann S (1994) Chem Mater 6:311
61. Litvin AL, Valiyaveettil S, Kaplan DL, Mann S (1997) Adv Mater 9:124
62. Kitano Y (1962) Bull Chem Soc Jpn 35:1973
63. Wang L, Sondi I, Matijević E (1999) J Colloid Int Sci 218:545
64. Richter A, Petzold D, Hofman H, Ullrich B (1995) Chem Tech 6:306
65. Chiroff RT, White RA, White EW, Weber JN, Roy DM (1977) J Biomed Mater Res 11:165
66. Stupp SI, Braun PV (1997) Science 277:1242
67. Wang RZ, Addadi L, Weiner S (1997) Philos Trans R Soc Rondon Ser B-Biol Sci 352:469
68. Berman A, Hanson J, Leiserowitz L, Koetzle TF, Weiner S, Addadi L (1993) Science 259:776
69. Phillips SE, Frisia S, Jones B, Van der Borg K (2000) J Sediment Res 70:1171
70. Olszta MJ, Gajjeraman S, Kaufman M, Gower LB (2004) Chem Mater 16:2355
71. Cölfen H, Mann S (2003) Angew Chem 42:2350
72. Weiss IM, Tuross N, Addadi L, Weiner S (2002) J Exp Zool 293:478
73. Hasse B, Ehrenberg H, Marxen JC, Becker W, Epple M (2000) Chem Eur J 6:3679
74. Lowenstam HA, Weiner S (1985) Science 227:51
75. Beniash E, Aizenberg J, Addadi L, Weiner S (1997) Proc R Soc London B 264:461
76. Li M, Mann S (2002) Adv Funct Mater 12:773
77. Kawano J, Shimobayashi N, Kitamura M, Shinoda K, Aikawa N (2002) J Crystal Growth 237:419
78. Perić J, Vučak M, Krstrulović R, Brečević L, Krali D (1996) Thermochim Acta 277:175
79. Berndt ME, Seyfried WE Jr (1999) Geochim Cosmochim Ac 63:373
80. Xu G, Yao N, Aksay IA, Groves JT (1998) J Am Chem Soc 120:11977
81. Lahiri J, Xu G, Dabbs DM, Yao N, Aksay IA, Groves JT (1997) J Am Chem Soc 119:5449
82. Zhang LJ, Liu HG, Feng XS, Zhang RJ, Zhang L, Mu YD, Hao JC, Qian DJ, Lou YF (2004) Langmuir 20:2243
83. Mann S, Didymus JM, Sanderson NP, Heywood BR, Samper EJA (1990) J Chem Soc Faraday Trans 86:1873
84. Geffroy C, Foissy A, Persello J, Cabane B (1999) J Colloid Inter Sci 211:45
85. Ogino T, Tsunashima N, Suzuki T, Sakaguchi M, Sawada K (1988) Nippon Kagaku Kaishi 6:899
86. Keum DK, Kim KM, Naka K, Chujo Y (2002) J Mater Chem 12:2449
87. Levi Y, Albeck S, Brack A, Weiner S, Addadi L (1998) Chem Eur J 4:389

88. Donners JJJM, Heywood BR, Meijer EW, Nolte RJM, Roman APHJ, Schenning APHJ, Sommerdijk NAJM (2000) Chem Commun, p 1937
89. Donners JJJM, Heywood BR, Meijer EW, Nolte RJM, Sommerdijk NAJM (2002) Chem Eur J 8:2561
90. Förster S, Antonietti M (1998) Adv Mater 10:195
91. Mann S, Archibald DD, Didymus JM, Douglas T, Heywood BR, Meldrum FC, Reeves NJ (1993) Science 261:1286
92. Mann S, Hannington JP, Williams RJP (1986) Nature 398:565
93. Meyer M, Wallberg C, Kurihara K, Fendler JH (1984) J Chem Soc Chem Commun, p 90
94. Yong H, Coombs N, Ozin GA (1997) Nature 386:692
95. Verdoes D, Kashichiev D, van Rosmalen GM (1992) J Crystal Growth 118:401
96. Boggavarapu S, Chang J, Calvert P (2000) Mater Sci Eng C11:47
97. Naka K, Keum DK, Tanaka Y, Chujo Y (2004) Bull Chem Soc Jpn 77:827
98. Balz M, Therese HA, Li J, Gutmann JS, Kappl M, Nasdala L, Hofmeister W, Butt HJ, Tremel W (2005) Adv Funct Mater 15:681
99. Naka K, Keum DK, Tanaka Y, Chujo Y (2000) Chem Commun, p 1537
100. Sugihara H, Ono K, Adachi K, Setoguchi Y, Ishihara T, Takita Y (1996) J Ceram Soc Jpn 104:832
101. Wie H, Shen Q, Zhao Y, Wang DJ, Xu DF (2003) J Cryst Growth 250:516
102. Keum DK, Naka K, Chujo Y (2003) Bull Chem Soc Jpn 76:1687
103. Keum DK, Naka K, Chujo Y (2004) J Cryst Growth 270:655
104. Kumar GS, Neckers DC (1989) Chem Rev 89:1915
105. Bullock DJW, Cumper CWN, Vogel AI (1965) J Chem Soc 5316
106. Menzel H (1994) Macromol Chem Phys 195:3747
107. Fissi A, Pieroni O (1989) Macromolecules 22:1115
108. Dante S, Advincula R, Frank CW, Stroeve P (1999) Langmuir 15:193
109. Natansohn A, Rochon P (2002) Adv Mater 14:869
110. Clavier G, Ilhan F, Rotello VM (2000) Macromolecules 33:9173
111. Keum DK, Naka K, Chujo Y (2004) Chem Lett 33:310

Inorganic–Organic Calcium Carbonate Composite of Synthetic Polymer Ligands with an Intramolecular NH···O Hydrogen Bond

Norikazu Ueyama (✉) · Kazuyuki Takahashi · Akira Onoda · Taka-aki Okamura · Hitoshi Yamamoto

Department of Macromolecular Science, Graduate School of Science, Osaka University, Toyonaka, 560-0043 Osaka, Japan
ueyama@chem.sci.osaka-u.ac.jp

1	Introduction	156
2	pK_a Shift by Neighboring Amide NH Through a Hydrogen Bond	161
3	Increase of Formation Constant by pK_a Shift	162
4	Partially Covalent Ca–O Bond in Carboxylate and Phosphate Ca(II) Complexes	164
5	Weak NH···O Hydrogen Bond in Sulfonate Ca(II) Complexes	166
6	Strong NH···O Hydrogen Bond in Carboxylate Ca(II) Complexes	167
7	Structural Transformation by Rearrangement of Hydrogen Bond Networks	169
8	Ca Cluster with Synthetic Chelating Ligand	173
9	$CaCO_3$/Polycarboxylate Composites as Models of Biominerals	174
10	$CaCO_3$ Composites with NH···O Hydrogen Bonds in the Polymer Main Chain	175
11	Location of Strongly Binding Polycarboxylate on the Surface of a $CaCO_3$ Crystal	176
12	Dependence of the $CaCO_3$ Composite on Polymer Ligand Tacticity	180
13	Biological Relevance of Synthetic Polymer Ligands	182
14	Nanocomposite of the Nacreous Layer in *Pinctada Fucata*	183
15	Variation of Metal–O Bonding by Conformational Change of Carboxylate Ligands	184
16	Proton-Driven Conformation Switch in Asp-Oligopeptide and Model Compounds	185

17	Binding Regulation of Polycarboxylate on CaCO$_3$ Crystals by Conformational Change 187
18	Conclusions 188
References	 189

Abstract Amide NH prelocating in the vicinity of a carboxylic acid group shifts the pK_a of carboxylic acid to a lower value. A carboxylate anion is stabilized by NH $\cdots$ O hydrogen bond under hydrophobic conditions. In the case of Ca carboxylate complex, the decrease of the pK_a value increases a formation constant involving the protonation process of carboxylic acid. In addition, the NH $\cdots$ O hydrogen bond controls the covalent character of the Ca–O bond for a totally neutral complex that possesses a shortening Ca–O bond distance. Both factors contribute to a strong binding of polycarboxylate on the surface of CaCO$_3$ crystals. The polycarboxylate ligands with a NH $\cdots$ O hydrogen bond give nanosized vaterite crystals and are located on the surface of the vaterite agglomerate. The deprotonation of some carboxylic acids leads to conformational change in Asp-X-Gly tripeptides, partially amidated Kemp's acid and monoamidated maleate. Conformational switching between stabilized and activated forms for carboxylate affords an unusual crystal growth of calcite even if one of the conformers has a strong NH $\cdots$ O hydrogen bond.

Keywords Ca complexes $\cdot$ Calcium carbonate $\cdot$ NH $\cdots$ O hydrogen bond $\cdot$ pK_a shift $\cdot$ Polycarboxylates $\cdot$ Prelocated amide NH

1
Introduction

Biomineralization of inorganic-organic materials is often found in biological systems [1–9]. Minerals comprise a large, diverse group, of which over 60 are known including Mg, Si, Ca, Mn, and Fe ions [10]. For example, bone and teeth, or shell and pearl are composed of Ca(PO$_4$)$_2$(OH)$_2$ or CaCO$_3$ crystals, respectively, in the combination of organic polymer ligands such as proteins, polysaccharides and lipids. The biomineral structure and the morphology of CaCO$_3$ minerals have been found to be remarkably diverse. The nacreous layer of *Pinctada fucata* pearl oyster is composed of aragonite crystals and a small amount of proteins; however, other biological CaCO$_3$ crystals have been observed to exhibit four morphologies (i.e., calcite, vaterite, aragonite, and amorphous). The *c*-axes of aragonite crystals in pearl are aligned in the same direction [11]. A highly acidic polypeptide having a repeated Gly-Xaa-Asn fragment (Xaa = Glu, Asp, or Asn), was isolated from the nacreous layer of *Pinctada fucata*. This protein is thought to be involved in the controlled crystallization of the aragonite phase, as schematically shown in Fig. 1 [12]. Coccolithophores cover their cell surface with coccoliths that are composed of three-dimensionally ordered calcite crystals [13]. An extensively acidic polysaccharide, the molecular weight of which is about ten thousand, was

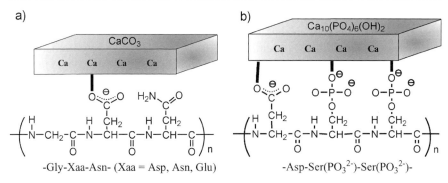

Fig. 1 Schematic drawings of **a** CaCO$_3$/protein composite in nacreous layer of pearl and **b** hydroxyapatite/protein composite in bone. CaCO$_3$ and apatite composites are proposed to contain -Gly-Xaa-Asn- (Xaa = Asp, Asn, Glu) and -Asp-Ser(PO$_3$H$_2$)-Ser(PO$_3$H$_2$)- fragments, respectively

isolated from the coccolith of both *Emiliania huxleye* [14] and *Pleurochrysis carterae* [15].

Bone and teeth are composed of hydroxyapatite crystals [16, 17] and the crystal axis of both is controlled by organic matrices, such as collagens, acidic peptides and peptide-glucans [18–20]. The acidic peptide is called phosphoryn and contains repeated Asp-Ser-Ser sequences, of which the Ser residues are totally phosphorylated (Fig. 1) [21]. Similarly, large numbers of Glu, Asp and Ser residues are found in peptide-glucan and dentin sialoprotein. These proteins are known to be essential for the calcification of hydroxyapatite crystals [18].

These Ca composites are attractive synthetic targets as an ecological architecture. The control of biomineralization involves compositional, structural, morphological, and organizational specificities in the organic phase [22, 23]. For example, a composition containing divalent metal ions affects the polymorph [24]. The structural specificity is often observed depending on the kind of organisms that utilize CaCO$_3$ to restrict the crystal polymorph. Morphological specificity is demonstrated in the specific habits of particles of inorganic materials in many organisms, e.g. pearl oyster. The extracted biopolymer causes the nucleation and growth of aragonite on calcite seed crystals [25]. Organization specificity is realized by the remarkable macrostructures produced by biomineralization, such as bone and shell. The organic macromolecular matrix is packed into the mineralization space. The organization and surface chemical properties of the macromolecular matrix contribute to the site-directed nucleation and mineral growth. Biopolymer ligands involved in the biominerals have been thought to bind strongly to the edge of inorganic crystals, although some organic matrices control both the mineralization and the Ca-dissociation of the crystals [5].

In general, the crystallization process removes the small amount of impurity in the system from the pure inorganic crystals. However, the biopolymer ligands are tightly ligated to the crystals to form organic–inorganic composites. Thus, the property characteristics of the strongly binding polycarboxylate to the surface of $CaCO_3$ seem to be crucial for biomineralization, accompanied by a carboxylate-orientated polymer matrix fixed to the crystal lattice. Many reviews have discussed how matrices control polymorphs during mineralization [26–31]. In addition, coordination information on synthetic Ca carboxylate complexes is important for the elucidation of the Ca – O bond on the surface of Ca biominerals.

These polymer ligands also play a significant role in the dissociation of the Ca(II) ion from $CaCO_3$ crystals in some biological systems. For example, crayfish store calcium ions in the stomach as amorphous $CaCO_3$, called gastroliths, just before their ecdysis. After ecdysis the stored $CaCO_3$ is transferred to the exoskeletons for reinforcement. Gastroliths contain chitins and proteins, which have Glu-rich [32, 33] and Gly-Ser-X-X-Phe peptide fragments [34]. The extended X-ray absorption fine structure analysis of lobster carapaces and cystolith-bearing plants indicates that the Ca site of amorphous $CaCO_3$ is unsymmetrically surrounded by six oxygen atoms [35]. The organic matrix seems to be involved in the regulation of Ca ion transport. The special polymer matrix and Mg(II) in the ascidian ion contribute to the stabilization of the intrinsically unstable amorphous $CaCO_3$ [36]. Similar phenomena are also observed in bone and were reported in the presence of an inhibitor protein during the mineralization of hydroxyapatite [18]. Biopolymer ligands play a crucial role in both mineralization and demineralization, albeit with opposite functions.

Although intensive efforts have been devoted to elucidating the correlation between the function and the structure of natural protein ligands, only a few studies have been conducted in this area because of the difficulty associated with analyzing biopolymer ligands in insoluble biominerals. Statherin is a multifunctional protein composed of 43 amino acid residues, inhibits the primary and secondary precipitation of hydroxyapatite in saliva and serves as a boundary lubricant [37]. The acidic N-terminal region of statherin is essential for its binding to hydroxyapatite [38]. A $T_{1\rho}$ measurement has been performed to analyze the main chain carbonyl groups in the 15-amino acid peptide of the *N*-terminus on binding to hydroxyapatite [39–41]. The binding region in the *N*-terminus gives a rigid dynamic and the following α-helical domain displays increased dynamics, indicating that the acidic region is involved in the binding of statherin to the hydroxyapatite. The regulation of mineralization is presumably associated with the conformational changes occurring in the biopolymer; however, there is a lack of detailed information about the structure of the biomineralizing protein.

To investigate the role of the polymer ligand and $CaCO_3$ for biomineralization, studies have examined several important areas, including calcinations by biopolymers, the construction of an organic-inorganic composite as a nanomaterial, the restriction of morphology upon $CaCO_3$ crystallization, and the binding properties of polymer ligand on the surface of CaCO$_3$ crystals. In this review, we focus on the strong Ca-binding on the surface of CaCO$_3$/polycarboxylate composite by emphasizing the function of the intramolecular NH $\cdots$ O hydrogen bond between the neighboring amide NH and the carboxylate oxyanion in the synthetic polymer ligands. In addition, a relationship between the restriction of the morphology of CaCO$_3$ crystallization and its strong binding has been studied using various types of synthetic polymer ligands containing carboxylate groups.

The chemical functions of the intramolecular NH $\cdots$ S hydrogen bonds between thiolate and neighboring amide NH have been elucidated using simple thiolate and Cys-containing peptide complexes. The hydrogen bond has several clear functions because the thiolate sulfur $p\pi$ orbital lobe is remarkably large compared with that of carboxylate oxygen. The presence of the NH $\cdots$ S hydrogen bond in mononuclear thiolate-metal complexes including Cu(I) [42], Mo(IV) [43], Cd(II), Fe(II) [44], and Co(II) ions [45], decreases the M – S bond distances when compared with those of the corresponding thiophenolate complexes without the hydrogen bond. For example, a Co(II)-thiolate complex has a tetrahedral structure with strong intramolecular NH $\cdots$ S hydrogen bonds (Fig. 2a). The Cd – S bond distance exhibited in [Cd^{II}(S-t-BuCONHC$_6$H$_4$)$_4$]$^{2-}$ exhibits a short Cd – S bond distance, compared with that of [Cd^{II}(S – C$_6$H$_5$)$_4$]$^{2-}$. The NH $\cdots$ S hydrogen bond decreases the sulfur $p\pi$ electron density in the highest occupied molecular orbital (HOMO) to weaken the related Cd – S antibonding $p\pi$–$d\pi$ interaction (Fig. 2b). Because of similarities in their electronegativities, the covalency in Mo – Se of [Mo^{IV}O(Se-2-t-BuCONHC$_6$H$_4$)$_4$]$^{2-}$ was found to have the same Mo – X distance as that of [Mo^{IV}O(S-2-t-BuCONHC$_6$H$_4$)$_4$]$^{2-}$. A clear COSY cross-peak is observed in the ^{77}Se – ^{1}H COSY spectrum with 5.4 Hz for $J(^{77}$Se – ^{1}H). The covalency is approximately 10% of the Se – H single bond in CH$_3$SeH ($J(^{77}$Se – ^{1}H), 41.7 Hz) (Okamura et al., personal communication).

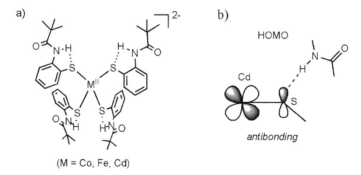

(M = Co, Fe, Cd)

Fig. 2 Molecular structure of **a** [M^{II}(S-t-BuCONHC$_6$H$_4$)$_4$]$^{2-}$ (M = Co, Fe, Cd) with intramolecular NH $\cdots$ S hydrogen bond and **b** *antibonding* interaction of Cd – S in HOMO of [Cd^{II}(S-t-BuCONHC$_6$H$_4$)$_4$]$^{2-}$ affected by amide NH

In contrast, the Fe – S bond distance in P450 model porphyrin complexes elongates in the presence of the NH $\cdots$ S hydrogen bond as well as that observed in a similar Ga(III) complex [46, 47]. Thus, the NH $\cdots$ S hydrogen bond controls the properties of the M – S bond, which affects the reactivity of a *trans*-coordinating substrate through *trans*-influence [48]. This type of regulation is important in terms of its effect on the stabilization and activation of the metal center in metalloproteins, metalloenzymes, and their model complexes. In the transition-metal complexes, the NH $\cdots$ S hydrogen bond shifts the redox potential toward the positive side [43, 46, 49, 50]. The presence of the hydrogen bond leads to the easy reduction of these electron-rich metal-thiolate complexes by a mild reductant.

Furthermore, P450 porphinate model complexes with cysteine-containing α-helical peptide have shown that properties of Fe – S bonds can be regulated by the NH $\cdots$ S bonds supported by the hairpin turn conformation and the following NH $\cdots$ O = C hydrogen bond networks in the helix [51].

Hydrogen bonds affect metal–oxygen bonds as well as the metal-thiolate complexes [52]. The NH $\cdots$ O hydrogen bond, which is similar to the above-mentioned NH $\cdots$ S hydrogen bond, has been found in metal complexes of phenolate with neighboring amide groups. An X-ray crystallographic analysis of beef liver catalase indicates that the axial ligand of heme is a tyrosine phenolate and that the axially ligated oxygen of the tyrosine residue has double NH $\cdots$ O hydrogen bonds with a neighboring arginine residue fixed by hydrogen bonding networks [53, 54], although the function is not clear. The crystal structures of [Fe^{III}(TPP){O-2,6-(CF$_3$CONH)$_2$C$_6$H$_3$}] (TPP = teraphenylporphinato) and [Fe^{III}(TPP)(O-2-CF$_3$CONHC$_6$H$_4$)], which were synthesized as models of catalase, predict the presence of NH $\cdots$ O hydrogen that contribute to the positive shift of the redox potential in Fe(III)/Fe(II). In this case, the Fe – O bond distance of [Fe^{III}(TPP)(O-2-CF$_3$CONHC$_6$H$_4$)] is similar to that of [Fe^{III}(TPP)(OPh)] [52]. OEP-Fe(III) complexes (OEP = octaethylporphinato) with these ligands at the axial position, [Fe^{III}(OEP)(O-2-CF$_3$CONHC$_6$H$_4$)], [Fe^{III}(OEP)(O-2,6-(CF$_3$CONH)$_2$C$_6$H$_3$)], [Fe^{III}(OEP)(OPh)] and [Fe^{III}(OEP)(O-2,6-(*i*-Pr)$_2$C$_6$H$_3$)], were synthesized, as compared with the corresponding TPP-Fe(III) complexes [55]. The NH $\cdots$ O hydrogen bond elongates the Fe – O bond distances and widens the Fe – O – C bond angles when compared with those in [Fe^{III}(OEP)(OPh)] without the hydrogen bond. The perturbation caused by the hydrogen bond towards the sulfur atom, which is presumably associated with a bonding Fe – O HOMO, weakens the Fe – O bond with the decrease of $d\pi$–$p\pi$ overlap.

The formation and the dissociation of calcium-oxygen bonds are known to be precisely controlled by calcium-binding proteins participating in the regulation of Ca concentrations in the biological system. For example, troponin C is involved in muscle contraction and the calcium-binding site is located in the loop region of the EF-hand structure [56]. Aspartate and glutamate residues are observed to bind to calcium ion and the oxygen atoms of

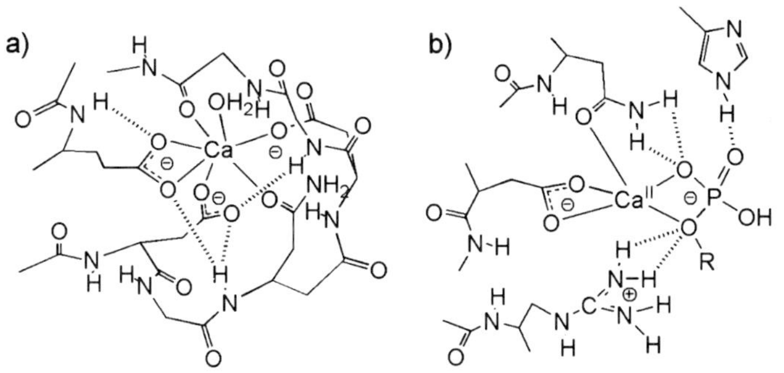

Fig. 3 Structures of **a** Ca-binding site in troponin C and **b** Ca center of mammalian phosphoinositide

these residues almost form hydrogen bonds with the main chain amide NHs (Fig. 3a). Similarly, NH $\cdots$ O hydrogen bonds are often found in phosphate or sulfate groups coordinating to metal ions in metalloproteins. In the active site of mammalian phosphoinositide-specific phospholipase C, the phosphate groups coordinating to Ca ion have several NH $\cdots$ O hydrogen bonds with surrounding peptide NHs (Fig. 3b) [57]. The phosphate group coordinating to adjacent Ca ions in deoxyribonuclease I also forms hydrogen bonds with His NHs [58, 59].

Sulfonate anion interacts with metal ions in biological proteins, and the sulfonate oxygen often forms NH $\cdots$ O hydrogen bonds [60]. In arylsulfatase B proteins, the sulfate anion is tightly bound, primarily by seven hydrogen bonds; five of which are donated by the main chain NH groups [61, 62]. The results obtained from various mutations of these proteins suggest that the NH $\cdots$ O hydrogen bonds to the coordinating oxo acid groups are functionally essential, especially as a hydrogen bond network. The hydrogen bond network contributes to the rigid folding of the protein chain.

2
pK_a Shift by Neighboring Amide NH Through a Hydrogen Bond

Thiol, phenol, phosphate monoanion and carboxylate exist as acids under hydrophobic and neutral conditions. Their deprotonated anion states coordinate to metal ions. The extent of this deprotonation depends on the pK_a value of these acids, because their deprotonation competes with that of coexisting water.

The functional role of the NH $\cdots$ O hydrogen bonds on the pK_a shift was examined in an aqueous micellar solution using bulky thiol, phenol [63], and carboxylic acid derivatives [64]. In general, pK_a values are obtained in an aqueous solution. However, using an aqueous micellar solution allows us to measure pK_a values even under the hydrophobic conditions, which can maintain intramolecular NH$\cdots$O hydrogen bonds accompanied by an adequate rapid rate of proton transfer. Intramolecularly prelocated amide NH shifts the pK_a values of these acids. Singly and doubly hydrogen-bonded benzenethiol, phenol and benzoic acid derivatives show a clear pK_a shift. The difference in pK_a values between 2-acylamino benzoic acid and 2,4,6-trimethyl benzoic acid is caused by the presence of neighboring amide NH rather than the stabilization of the anion state with hydrogen bonds.

3
Increase of Formation Constant by pK_a Shift

The pK_a value reflects the formation and dissociation constants of metal-carboxylate bonds. The dissociation is prevented by the presence of NH$\cdots$O hydrogen bonds of coordinating carboxylate because the NH$\cdots$O hydrogen bonds lower the pK_a value of the corresponding carboxylic acid [65–67]. The hydrogen bonds toward the coordinating oxygen or sulfur atom of phenolate [63] or thiolate have a similar function with each other [43, 68].

The conventional formation constant K, as shown in Eq. 1, is the most widely accepted one for the scale determining the strength of Ca-ligand binding as shown in Fig. 4a. This equilibrium is true only under the assumption that a carboxylate anion interacts with one Ca^{2+} ion. The strength of Ca–O bonds in Ca complexes was examined using a competitive ligand exchange reaction between Ca-carboxylate and carboxylic acid [69]. During the synthesis of hydrogen-bonded Ca complexes, an NH$\cdots$O hydrogen-bonded benzoic acid extracts Ca ion from a Ca complex of nonsubstituted benzoate. As shown in Fig. 4b, the most appropriate index for measuring the strength of the weak acid-metal interaction involving a deprotonation process is the formation constant, K_{HL} (defined in Eqs. 2 and 3).

$$K = \frac{[Ca(OCOAr)^+]}{[Ca]^+ [ArCOO^-]} \tag{1}$$

$$K_{HL} = \frac{[Ca(OCOAr)^+][H^+]}{[Ca^{2+}][ArCOOH]}$$

$$= \left\{ \frac{[Ca(OCOAr)^+]}{[Ca^{2+}][ArCOO^-]} \right\} \left\{ \frac{[H^+][ArCOO^-]}{[ArCOOH]} \right\}$$

$$= K \times K_a \tag{2}$$

$$\log K_{HL} = \log K - pK_a . \tag{3}$$

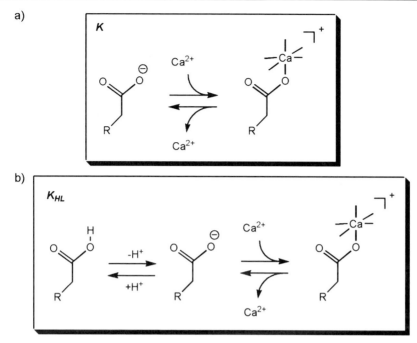

Fig. 4 a Conventional complex formation constant (K) between deprotonated carboxylate and Ca ion and **b** the alternative complex formation constant (K_{HL}) including a high energy barrier in deprotonation process

The stabilization of the metal–oxygen bond by the NH $\cdots$ O hydrogen bond was established for a Tb(III) complex, the ionic radius of which is similar to that of Ca(II) [70]. A Tb(III) complex of 2,6-bis(acetylamino)benzoate exhibits a higher emission intensity than that of nonsubstituted benzoate in aqueous solution. Generally, the Tb^{3+} ion exists as an aqua complex, $[Tb(OH_2)_9]^{3+}$, in the absence of anion ligands [71, 72]. These coordinated water molecules serve as efficient quenchers of the intrinsic Tb(III) luminescence; the emission intensity of an aqueous terbium solution is normally very weak under these conditions. In the presence of 2,6-bis(acetylamino)benzoate ligand, the carboxylate ligand forms the stable Tb(III) complex assisted by a NH $\cdots$ O hydrogen bond and displaces water molecules. The formation of the NH $\cdots$ O hydrogen bond stabilizes the metal–oxygen bond that is protected from dissociation by hydrolysis. Thus, the NH $\cdots$ O hydrogen bond contributes to the strong binding of an anion ligand to the Ca ion.

Various metal complexes of carboxylate with intramolecular NH $\cdots$ O hydrogen bonds can be readily synthesized [65, 67]. The carboxylate ligand with the hydrogen bond promotes a ligand exchange reaction as shown in Eq. 4, which proceeds quantitatively because of the difference in pK_a value between the two carboxylic acids.

$$M^{II}(OCO - 2,4,6 - {}^iPr_3Ph_n(OH_2)_{4-n} + ArCOOH \rightarrow$$
$$M^{II}(OCOAr)_n(OH_2)_{4-n} + 2,4,6 - {}^iPrPhCOOH$$

$(M = Ca, Zn, Cd; Ar = 2,6\text{-}(t\text{-BuCONH})_2C_6H_3COO^-)$. (4)

Thus, Ca – O bonds between biominerals and biopolymers can also be controlled by the hydrogen bond networks that are formed with polypeptides. Regulation of Ca – O bonds in biomineral systems is assumed to be achieved by transforming the peptide structure that is correlated with rearrangement of the hydrogen bond network. A series of Ca complexes with oxo acids, such as phosphate, sulfonate, and carboxylate ligands, that possess strategically oriented bulky amide groups have been studied to elucidate the roles of NH···O hydrogen bonds as models of the networks (Onoda et al., personal communication) [73]. Novel analogues with the hydrogen bonding networks in calcium clusters with designed bulky amide ligands were synthesized to investigate the transformation of the calcium clusters cooperating with the reorganization of the hydrogen bonding networks [74].

4
Partially Covalent Ca – O Bond in Carboxylate and Phosphate Ca(II) Complexes

Various Ca(II) complexes with carboxylates have been synthesized and crystallographically characterized [75]. Three modes are most commonly observed for these complexes; the unidentate mode, the bidentate mode, and the *anti*-mode. In the unidentate mode, the Ca ion interacts with only one of the two O atoms of the carboxylate group. The coordinating atom is a carboxylate oxyanion rather than the carbonyl O atom. In the bidentate mode, the two O atoms of the carboxylate chelates to one Ca ion. In the *anti*-mode, the Ca complex is chelated between one of the carboxylate O atoms and a neighboring atom. Furthermore, the *anti*-mode is involved in a polynuclear mode with the O atom of the carboxylate forming a bridge between two Ca ions.

The difference between the unidentate and bidentate modes is based on the type of Ca – O – C angle. The bidentate complexes display an acute Ca – O – C angle in the range of 80° ~ 100° whereas the unidentate complexes exhibit a relatively obtuse angle in the range of 110° ~ 180°. A relationship between the Ca – O bond distances and the Ca – O – C angles has been reported for various carboxylate Ca complexes as illustrated in Fig. 5a [75]. The average Ca – O bond distance is 2.38 Å, within 2.3 ~ 2.5 Å. We reported a similar relationship between Ca – O bond distances and Ca – O – P angles in various Ca complexes with O_2POR, $O_2P(OH)R$, O_3PR, $O_2P(OH)R$ ligands (Fig. 5b) (Onoda et al., personal communication). Both scatter plots indicate a tendency

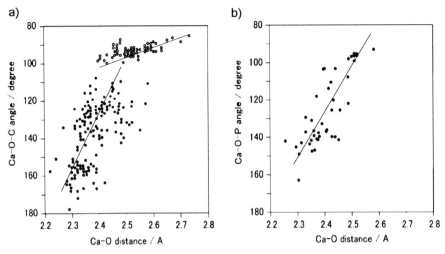

Fig. 5 Scatter plots of Ca – O bond distances against Ca – O – P bond angles in various **a** Ca(II)-carboxylate [138] and **b** Ca(II)-phosphate complexes [85, 95, 110–113, 139]. The *circles* and *squares* refer to unidentate and bidentate modes, respectively

for the Ca – O distance to be shorter when the Ca – O – C and Ca – O – P angles become larger, thus showing a roughly collinear relationship. It is likely that the shortness of Ca – O bond distance, accompanied by an increase in the Ca – O – C or Ca – O – P angle, is attributed to the π-interaction between the $d\pi$ or $p\pi$-orbitals of Ca(II) and the $p\pi$-orbitals of the oxygen atoms in the Ca – O bond. Therefore, the Ca – O bond in the carboxylate and phosphate Ca complexes is believed to have a somewhat covalent character.

A tetrakis carboxylate anion Ca complex, $[Ca\{OCO-2,6-(t-BuCONH)_2C_6H_3\}_4]^{2-}$, exhibits a regular Ca – O bond distance of 2.4 Å and an acute Ca – O – C angle with a bidentate mode (Fig. 6a). Thus, a totally anionic Ca complex has a relatively long Ca – O bond distance, presumably with a weak Ca – O bond due to a typical cationic bond character. On the other hand, a neutral Ca(II) complex, $Ca\{OCO-2,6-(CH_3CONH)_2C_6H_3\}_2(H_2O)_2$, exhibits a relatively shorter Ca – O bond distance of 2.278(2) Å with an obtuse Ca – O – C angle of 164.4(2)° in a unidentate mode (Fig. 6b). A theoretical analysis of the relationship between Ca – O bond distance and the Ca – O – C angle was carried out using ab initio molecular orbital calculations that took into consideration the d orbital (RHF/6-311+G** base set for Ca, RHF/6-311++G** for C, H, N, and O). The results indicate that an obtuse Ca – O – C angle increases the d-orbital occupation number in Ca(II) to form a bonding $p\pi$–$d\pi$ interaction with oxygen $p\pi$ (Onoda et al., personal communication). Figure 6c,d illustrates the HOMO calculated for a neutral model complex, $[Ca(O_2CCH = CHNHCOCH_3)_2(CH_3CONH_2)_2(OH_2)_2]$. A clear bonding interaction of $p\pi$–$d\pi$ in the Ca – O bond is observed; how-

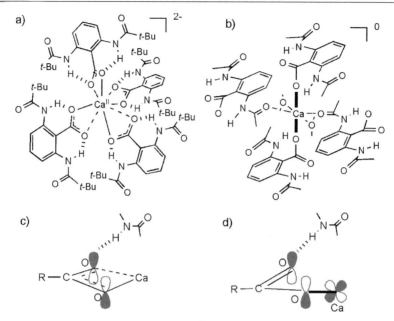

Fig. 6 Molecular structures of **a** [Ca^{II}{O$_2$C – C$_6$H$_3$-2,6-(NHCO-t-Bu)$_2$}$_4$]$^{2-}$ and **b** Ca^{II}{OCO-2,6-(CH$_3$CONH)$_2$C$_6$H$_3$}$_2$(H$_2$O)$_2$. Effect of the hydrogen bonds on **c** HOMO in [Ca^{II}{O$_2$C–C$_6$H$_3$-2,6-(NHCO-t-Bu)$_2$}$_4$]$^{2-}$ and **d** HOMO in Ca^{II}{OCO-2,6-(CH$_3$CONH)$_2$C$_6$H$_3$}$_2$(H$_2$O)$_2$

ever, an anion complex, [Ca(O$_2$CCH = CHNHCOCH$_3$)$_4$], does not have any $p\pi$–$d\pi$ interaction in the HOMO. The NH$\cdots$O hydrogen bond increases the bonding orbital in the Ca – O bond. It is emphasized that a neutral Ca complex has a shorter Ca – O bond distance, which corresponds to a strong Ca – O bond. A carboxylate ligand adopts two types of NH$\cdots$O hydrogen bonds, which compete with each other depending on the strength of the oxyanion charge. The covalent character of the Ca – O bond in the neutral complex, Ca{OCO-2,6-(CH$_3$CONH)$_2$C$_6$H$_3$}$_2$(H$_2$O)$_2$, also supports the strong binding of the carboxylate ligand to the Ca ion.

5
Weak NH$\cdots$O Hydrogen Bond in Sulfonate Ca(II) Complexes

Sulfonate Ca(II) complexes demonstrate a common ionic Ca – O bond character but do not display a correlation between Ca – O bond strength and oxyanion basicity. Intermolecular NH$\cdots$O hydrogen bonds to the sulfonate oxygen has been studied in the solid state and in solution [76–80]. The crystal structure of benzene sulfonate derivatives containing an amino group indicated the presence of intermolecular NH$\cdots$O hydrogen bonds from the ammonium

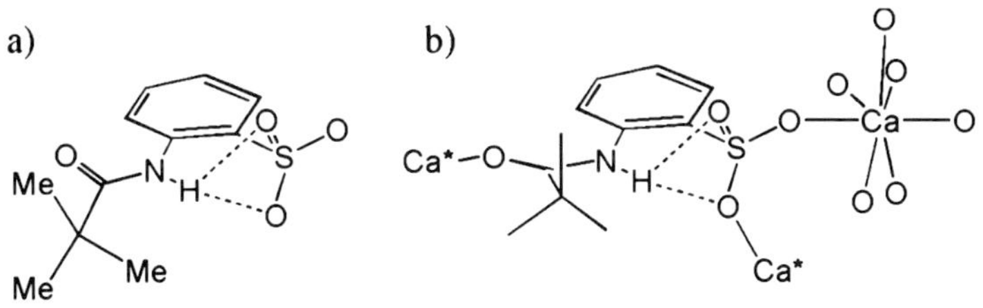

Fig. 7 Molecular structures of **a** (PPh$_4$)(SO$_3$-2-t-BuCONHC$_6$H$_4$) and **b** [Ca$_2$(SO$_3$-2-t-BuCONHC$_6$H$_4$)$_2$ (H$_2$O)$_4$]$_n^{n-}$

NH to the sulfonate oxygen. This kind of intermolecular NH $\cdots$ O hydrogen bond from the ammonium NH has often been used in crystal engineering. The molecular recognition between the sulfonate derivatives and the amide-containing acceptors has been discussed in terms of their binding constants during the formation of intermolecular NH $\cdots$ O hydrogen bonds [81, 82]. Intramolecular NH $\cdots$ O hydrogen bonds to the sulfonate oxygen, which have a low pK_a value, were studied using 2-acylaminobenzene sulfonic acid with the amide NH near the SO$_3$H group.

As supported by IR and ^{1}H NMR analyses, the molecular structures of the novel intramolecular NH $\cdots$ O hydrogen-bonded Ca arylsulfonate complex, [Ca$_2$(SO$_3$-2-t-BuCONHC$_6$H$_4$)$_2$(H$_2$O)$_4$]$_n$(2-t-BuCONHC$_6$H$_4$SO$_3$)$_{2n}$; the sulfonate anion, (HNEt$_3$)(SO$_3$-2-t-BuCONHC$_6$H$_4$), (PPh$_4$)(SO$_3$-2-t-BuCONH C$_6$H$_4$), (n-Bu$_4$N)(SO$_3$-2-t-BuCONHC$_6$H$_4$); and sulfonic acid, 2-t-BuCONHC$_6$H$_4$SO$_3$H, depict the presence of the formation of intramolecular NH $\cdots$ O hydrogen bonds between the amide NH and S – O, both in the solid state and in solution [73]. Sulfonic acid and the sulfonate anion and its Ca complex have a substantially weak (electrostatic) intramolecular NH $\cdots$ O hydrogen bond between the sulfonate oxygen and the amide NH. Compared with the structure of the sulfonate anion, (PPh$_4$)(SO$_3$-2-t-BuCONHC$_6$H$_4$) (Fig. 7), the amide NH in [Ca$_2$(SO$_3$-2-t-BuCONHC$_6$H$_4$)$_2$(H$_2$O)$_4$]$_n$(2-t-BuCONHC$_6$H$_4$SO$_3$)$_{2n}$ is not directed to one of the two sulfonate oxygen atoms but is located in the middle of two oxygen atoms (Fig. 7). It is likely that a weak NH $\cdots$ O hydrogen bond between the amide NH and the sulfonate oxygen is formed due to the strong conjugation of the sulfonate group, which results in lowering the basicity of the oxyanion. In this case, the Ca – O bond possesses an ionic character.

6
Strong NH $\cdots$ O Hydrogen Bond in Carboxylate Ca(II) Complexes

The amide NH ^{1}H NMR signals of an amide-derived benzoic acid ligand, 2,6-(t-BuCONH)$_2$C$_6$H$_3$COOH, its anion form, (NEt$_4$){2,6-(t-BuCONH)$_2$C$_6$H$_3$COO}, and its Ca complex, (NEt)$_4$[Ca^{II}\{OCO-C$_6$H$_3$-2,-6-(NHCO-t-Bu)$_2$\}$_4$]

monitor the strength of their NH $\cdots$ O hydrogen bonds in acetonitrile-d_3 solution. The amide NH ^{1}H signals of the carboxylate anion show a large shift of 2.25 ppm from that of the carboxylic acid, whereas the Ca complex exhibits a relatively large shift of 1.88 ppm. These findings indicate that the NH $\cdots$ O hydrogen bond between the amide NH and the carboxylate oxygen in the Ca complex is strong but that of the carboxylate anion form is weaker.

The behavior of the amide NH signals in acid and in anion states are absolutely different between sulfonic acid and benzoic acid, which have different pK_a values. The ability to form a hydrogen bond has been thought to be correlated with the basicity of the oxyanion which can be evaluated by using a ^{1}H NMR shift of the amide NH group as a diagnostic tool. The downfield shifts related to the deprotonation of sulfonic acid and carboxylic acid are 0.61 and 2.25 ppm, respectively. The large shift in the carboxylate anion form indicates that it has a significant ability to form a strong hydrogen bond, whereas the sulfonate anion is a weak hydrogen-bond acceptor due to the stronger conjugation in the sulfate anion. Carboxylic acid groups have relatively high pK_a values compared to those of sulfonic acid groups, and the charge is more localized in RCO_2^- than in RSO_3^-, thus making the carboxylate anion the stronger hydrogen-bond acceptor. The strength of each hydrogen bond in carboxylate, phosphate and sulfonate is represented in Fig. 8.

The ability of a hydrogen-bond acceptor of the oxyanion changes with Ca(II) coordination. The shift of the amide NH signals in the free anion and in the Ca complex results in a difference between the RSO_3^- and RCO_2^-. The downfield shift corresponds to the coordination of the Ca ion in RSO_3^- and is 0.07 ppm, whereas that in RCO_2^- is 1.45 ppm. The NH $\cdots$ O hydrogen bonds in the carboxylate anion in $(NEt_4)\{2,6-(t-BuCONH)_2C_6H_3COO\}$ are much stronger than those in the Ca carboxylate complex, $(NEt)_4[Ca^{II}\{OCO-C_6H_3-2,-6-(NHCO-t-Bu)_2\}_4]$. The strengths of the NH $\cdots$ O hydrogen bond in the sulfonate anion and in the Ca sulfonate complex are not substantially different. The decrease in the oxyanion basicity of the NH $\cdots$ O hydrogen bond in the Ca carboxylate complex is presumably caused by a charge transfer from the coordinated oxygen atom to the Ca ions. In contrast, RSO_3^-, which is a stable anion with strong conjugation, has fewer charge transfers occurring from the O(O – S) atoms to the Ca ions. The charge transfer from the oxyanion ligand to Ca(II) is associated with the $p\pi-d\pi$ interaction in the Ca – O bond.

The difference in the Ca – O bond between the carboxylate and the sulfonate Ca complexes is related to the regulatory properties of the NH $\cdots$ O hydrogen bonds. It has been proposed that the NH $\cdots$ O hydrogen bonds to the coordinated oxygen atoms can regulate the Ca – O bond by preventing the Ca – O bonds from dissociating as a result of a ligand exchange reaction caused by the lowering pK_a value of the ligands [52, 65]. The ^{1}H NMR results for the carboxylate Ca complexes indicate that the NH $\cdots$ O hydrogen bonds can decrease the basicity of the oxyanion which is correlated with the charge transfer from $RCOO^-$ to Ca ions.

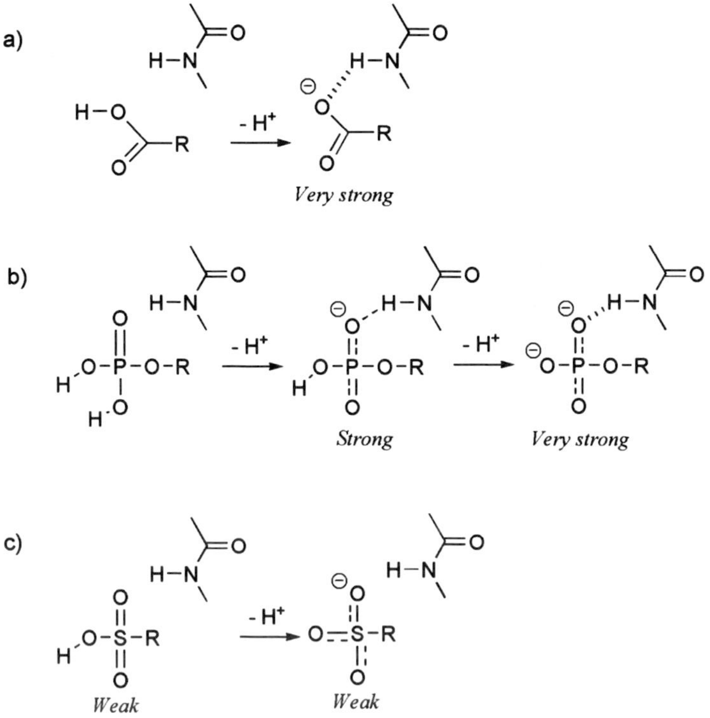

Fig. 8 Strength of NH⋯S hydrogen bond in **a** carboxylate, **b** phosphate and **c** sulfonate

The NH⋯O hydrogen bonds are weakly formed not only in the sulfonic acid state but also in the Ca sulfonate complex as described above. When comparing these two states, the difference in the strength of ionic Ca–O(sulfonate) bond character with the NH⋯O hydrogen bonds is considered minimal. Therefore, regulating the Ca–O bond character with the NH⋯O hydrogen bond is not available in the case of the highly ionic Ca–O bonds with sulfonate because of the strong delocalization occurring in RSO_3^-.

Thus, carboxylate and phosphate monoanion, which have a higher degree of basicity, possess hydrogen bonds with properties that differ from those found in sulfonate. A systematic investigation of the properties of the hydrogen bonds reveals that strong NH⋯O hydrogen bonds are formed only with O atoms of high basicity that coordinate to the metal ions.

7
Structural Transformation by Rearrangement of Hydrogen Bond Networks

All of the reported homoleptic Ca-carboxylate [83–93], phosphate [94–102] and sulfonate complexes [103–106] have polymeric structures. A correlation between the Ca – O bond and the NH···O hydrogen bonds can be discussed only in terms of the mononuclear structure of Ca complexes, because the Ca – O bond properties, including a bridging O atom, cannot be evaluated for the oxyanion basicity. Bulky carboxylate ligands enforcing the formation of a mononuclear core have been designed. The bulky amide group in the ligand works as a neutral hydrogen bond donor to regulate the strength of hydrogen bonding as precisely detected by ^{1}H NMR and IR spectroscopy.

Various polynuclear phosphate metal complexes with mainly intermolecular hydrogen bonds have provided a fascinating insight into the design of solid-state materials [100, 107, 108]. These Ca(II) phosphate complexes have been studied in terms of their biological relevance, especially understanding the biomineralization of Ca(II) phosphate materials, such as bone and teeth [18–20]. Many synthetic metal phosphate complexes with small ligands are known to have an open-framework structure, and then intermolecular hydrogen bonding networks contribute to the development of a variety of metal phosphate complexes as applied for ion exchange and catalyst [100, 107, 109]. For example, Ca(II) complexes with phosphate, ROPO$_3^{2-}$, and phosphonate ligands, RPO$_3^{2-}$, such as Ca(O$_3$POCH$_2$CH$_2$NH$_3$), have a one-dimensional structure [110, 111] and the others, such as Ca(O$_3$PMe), have layered structures [112, 113]. These small ligands predominantly form polymeric structures.

Extremely bulky and less bulky amide ligands can control the coordination geometry. Actually, a phosphate monoanion complex, (NMe$_4$)[Ca^{II}{O$_2$P(OH)OC$_6$H$_3$-2,6-(NHCOCPh$_3$)$_2$}$_3$(N≡CMe)$_3$], and a phosphate dianion complex, [Ca^{II}{O$_3$POC$_6$H$_3$-2,6-(NHCOCPh$_3$)$_2$}(H$_2$O)$_3$(MeOH)$_2$], which have an extremely bulky triphenylacylamino group, are forced to form a mononuclear Ca(II) core because of steric congestion (Fig. 9). NH···O hydrogen bonds

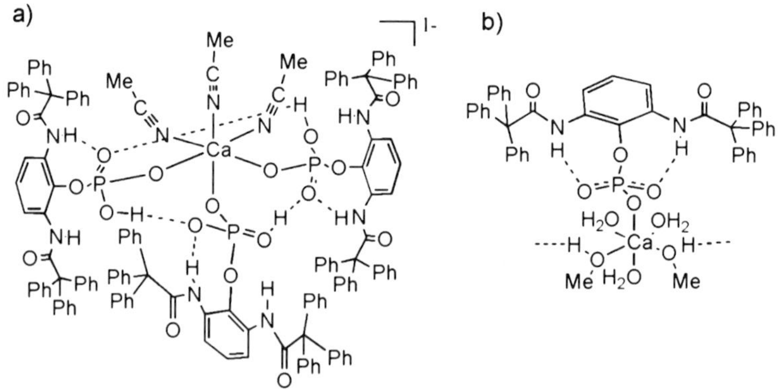

Fig. 9 Molecular structures of **a** (NMe$_4$)[Ca^{II}{O$_2$P(OH)OC$_6$H$_3$-2,6-(NHCOCPh$_3$)$_2$}$_3$(N≡CMe)$_3$] and **b** a phosphate dianion, [Ca^{II}{O$_3$POC$_6$H$_3$-2,6-(NHCOCPh$_3$)$_2$}(H$_2$O)$_3$(MeOH)$_2$]

are not formed in the phosphoric acid state, whereas the hydrogen bond is formed strongly in the monoanion state, and becomes very strong in the phosphate dianion state.

Less bulky ligands are able to restrict the coordination mode to the side of the Ca cluster. A less-bulky phosphoric acid, 2,6-$(PhCONH)_2C_6H_3OPO_3H_2$, gives three novel polynuclear Ca(II)- and Na(I)-phosphate complexes [74]. One is a zigzag-chain Ca cluster, $[Ca^{II}\{O_3POC_6H_3\text{-}2,6\text{-}(NHCOPh)_2\}(H_2O)_4(EtOH)]_n$, another a cyclic octanuclear form, $[Ca_8^{II}\{O_3POC_6H_3\text{-}2,6\text{-}(NHCOPh)_2\}_8(O=CHNMe_2)_8(H_2O)_{12}]$, and the last a hexanuclear complex, $(NHEt_3)[Na_3\{O_3POC_6H_3\text{-}2,6\text{-}(NHCOPh)_2\}_2(H_2O)(MeOH)_7]$. The crystallographic structures reveal that all have an *unsymmetric* ligand position because of the presence of the less bulky amide groups. The dynamic transformation of the zigzag chain Ca structure to the cyclic octanuclear Ca complex is induced by the addition of *N,N*-dimethylformamide (DMF) due to the coordination of DMF molecules (Fig. 10) [74]. This transformation is accompanied by a reorganization of the intermolecularly and intramolecularly hydrogen bonded networks. The DMF coordination breaks one of the hydrogen bonds, which results in the rearrangement of the network (Fig. 10).

The zigzag chain Ca cluster, the cyclic octanuclear Ca(II) complex and the hexanuclear Na(I) phosphate complex have unusual coordination geometry

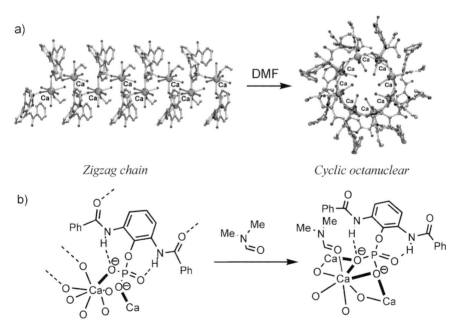

Fig. 10 a Transformation from Ca(II)-phosphate zigzag cluster to octanuclear cluster by DMF. **b** Rearrangement of hydrogen bond networks by the addition of DMF

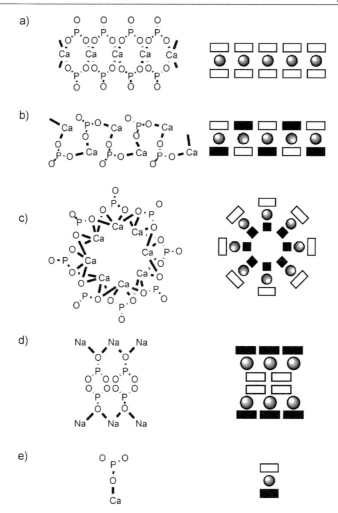

Fig. 11 A schematic representation of **a** a one-dimensional structure of reported simple phosphate Ca complexes, **b** a zigzag structure of less bulky phosphate Ca complex, **c** a cyclic octanuclear structure of the same less bulky phosphate Ca complex, **d** a hexanuclear structure of the less bulky phosphate Na(I) complex, and **e** a mononuclear structure of an extremely bulky phosphate Ca complex with unsymmetric ligand coordination. *White* and *black boxes* represent organic phosphate and other ligands (water and methanol), respectively. *Circle* refers to metal ion

with the hydrogen bond networks. Simple phosphate ligands also produce a polymeric Na cluster. Fig. 11 shows the schematic representation of various phosphate complexes.

Similarly, biopolymer and synthetic polymer ligands successively connected as the less-bulky phosphate and carboxylate ligands can precisely coordinate to the surface of Ca clusters, calcium carbonate or hydroxyapatite.

8
Ca Cluster with Synthetic Chelating Ligand

Orientation in the coordination of polycarboxylate and polyphosphate ligands to the one side of Ca clusters is important for biomineralization. The complexation advance by two less bulky ligands connected with a spacer was examined as models of Ca-cluster binding proteins. A dinuclear calcium-binding site was observed in thermolysin, as confirmed by crystallographic analysis [114], which also revealed the binding sites of two other calcium ions and one zinc ion. A similar dinuclear Ca site was proposed for the C2 domain of protein kinase C [115]. Three carboxylate ligands bridge between the two calcium ions in the dinuclear calcium site, which consists of 8-coordinated and 6-coordinated Ca ions. This dinuclear calcium-binding site has been discussed as a cooperative coordination of the two metal ions [116].

In calcium-binding proteins, hydrogen bond networks have often been found between the carboxylate oxygen atom and the main chain amide NH. Hydrogen bonding networks are believed to play a role in enabling the fixation of the carboxylate ligand orientation and in maintaining a suitable main chain conformation in the calcium-binding loop. Synthetic Ca carboxylate complexes revealed that the NH···O hydrogen bond protects the metal-carboxylate bond from dissociation with water due to the shift of pK_a of the conjugated carboxylic acid.

A synthetic dinuclear calcium model complex, [Ca_2{(2-OCO-3-$CH_3C_6H_3$NHCO$)_2$C(CH$_3$)$_2$}$_2$(CH$_3$OH)$_6$], was synthesized as a structural model of 8-coordinated Ca ions in the double calcium-binding site of thermolysin [65]. The complex has four NH···O hydrogen bonds between the amide NH and the carboxylate oxyanion (Fig. 12). Two types of bridging coordination of the carboxylate ligand to Ca(II) were found in the Ca(II) dimer complex. Of the two carboxylate oxygen atoms, the amide NH forms a strong NH···O hydrogen bond with the anionic one. The ligand exchange reac-

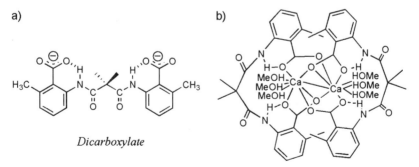

Fig. 12 Molecular structure of **a** dicarboxylate ligand and **b** dinuclear Ca complex, [Ca_2{(2-OCO-3-CH$_3$C$_6$H$_3$NHCO)$_2$C(CH$_3$)$_2$}$_2$(CH$_3$OH)$_6$]

tion between the dinuclear calcium complex and eight equimolar amounts of 2,4,6-trimethylbenzoic acid or 2-CH_3-6-t-$BuCONHC_6H_3COOH$ indicates that the NH $\cdots$ O hydrogen bond prevents the dissociation of the Ca – O bond. Although the COO plane is twisted from the benzene ring, the amide NH is directed toward one of the carboxylate oxygen atoms to form an NH $\cdots$ O hydrogen bond. The combined data indicate that a strong NH $\cdots$ O hydrogen bond is formed between the amide NH and the C – O$^-$ anion oxygen. The coordination of six methanol molecules outside the dinuclear site suggests a convergence of these water molecules in this type of Ca-binding complex by these bulky ligands. The chelating dicarboxylate ligand predominantly forms a neutral or anionic Ca complex in the presence of an anionic ligand by the extraction of the Ca ion from the surface of the $CaCO_3$ crystals in methanol/water. This reaction proceeds as follows (Eq. 5).

$$2CaCO_3 + 2\,[2\text{-}COOH\text{-}3\text{-}CH_3C_6H_3NHCO]_2\,C(CH_3)_2]_2] + 6CH_3OH$$
$$\rightarrow [Ca_2\{(2\text{-}OCO\text{-}3\text{-}CH_3C_6H_3NHCO)_2C(CH_3)_2\}_2(CH_3OH)_6]$$
$$+ CO_2 + 2H_2O \tag{5}$$

9
CaCO$_3$/Polycarboxylate Composites as Models of Biominerals

The environments around the Ca ion on the surface of $CaCO_3$ and hydroxyapatite crystals are quite unusual; however, pure crystals are still thought to contain some coordinated water molecules to compensate the charge. When one carboxylate coordinates to a Ca ion at the edge of a crystal, the neutral charge is still maintained because the twitter cation cannot exist on the surface.

A strategy for the synthesis of inorganic-organic materials has been adopted for the construction of micro crystals connected with a polymer ligand (e.g., a pearl brick). Various kinds of synthetic polymers and biopolymers possessing acidic substituents have also been examined as models of biominerals. Biopolymers with repeating oriented acidic groups seem to regulate the nucleation and growth of hierarchical nano-structures of the inorganic crystals of $CaCO_3$ and of the $Ca(PO_4)_2(OH)_2$ biominerals [3–7, 9].

Various artificial complexations consisting of synthetic polycarboxylates and peptides have been investigated as models of biomineralization. Examples of these synthetic polycarboxylates and peptides include the following ligands: poly(acrylate), poly(Glu) or poly(Asp) [117]; a stearic acid monolayer [9]; glycoproteins [118]; chitin fibers [119]; a porphyrin-template monolayer [120]; mercaptophenol-protected gold colloids [121]; dendrimer [28]; and chiral poly(isocyanide) [122]. These ligands have been thought to control the polymorph identified, as analyzed by IR or microscopy. The oriented crystallization of calcite, aragonite and vaterite has been demonstrated using a cross-linked collagenous matrix that includes poly(Asp) or poly(Glu) [123]. A matrix consisting of poly(acrylate) and cellulose has been used to form a $CaCO_3$ film in simulating the construction of a large pearl brick [124].

A similar control of the polymorph was performed by adding a radical initiator to a $CaCO_3$ solution containing sodium acrylate [125]. A thin film of calcite was produced using the monolayer assembly of porphyrin with carboxylate in the presence of poly(acrylate) at room temperature [120]. Furthermore, thin-films of aragonite were obtained in the presence of chitosan, resulting from the cooperation of poly(Asp) and $MgCl_2$ [119]. It is likely that poly(acrylate) and poly(Asp), which inhibit crystal growth, played a role in the successful formation of both films. A synthetic oligopeptide that was designed as a model of calcium-binding proteins, binds to the {110} faces of calcite. It has been proposed that the growth of calcite crystals is controlled by the α-helical conformation of the peptide ligand [126]. However, the location of these synthetic polymers, and even native biopolymers, on the crystal surface is still unclear because the large contents of solvent-insoluble crystalline $CaCO_3$ disturb the analysis of these polymer ligands by spectroscopic methods.

The ^{13}C solid-state NMR spectroscopy of the nacreous layer of *Pictada fucata* indicates a short T_1 relaxation time compared with synthetic Ca composites [127]. The nacreous layer also showed anomalous high efficiency of cross-polarization to the carbonate carbon in the aragonite during the cross-polarization/magic angle spinning (CP/MAS) experiments. These results indicate that the aragonite brick in the nacreous layer is not a simple single crystal, as previously predicted in the literature. The aragonite brick consists of nanoparticles of $CaCO_3$ as described below. Each nanoparticle has a similar crystal axis orientation in the aragonite brick; the orientation differs among the neighboring aragonite bricks.

10
$CaCO_3$ Composites with NH $\cdots$ O Hydrogen Bonds in the Polymer Main Chain

Three rigid polyamides containing carboxylic acid, $\{NHC_6H_3(COOH)NH COC(CH_3)_2CO\}_n$, $\{NHC_6H_3(COOH)NHCO\text{-}m\text{-}C_6H_4CO\}_n$, and $\{NHC_6H_3 (COOH)NHCOCH=CHCO\}_n$, are utilized as models of biomineralizing proteins (Fig. 13) [66]. All of the polymer ligands have an intramolecular 6-membered ring of NH $\cdots$ O hydrogen-bonded carboxylate groups that bind strongly to the Ca ion. Calcium carbonates/polymer ligand composites show that the former two ligands bind to the $CaCO_3$ cluster but that the latter polymer ligand is not involved in the $CaCO_3$ cluster. This finding indicates that the relative position of the carboxylate group in the polymer ligands primarily influences their ability to bind to the $CaCO_3$ cluster. The former

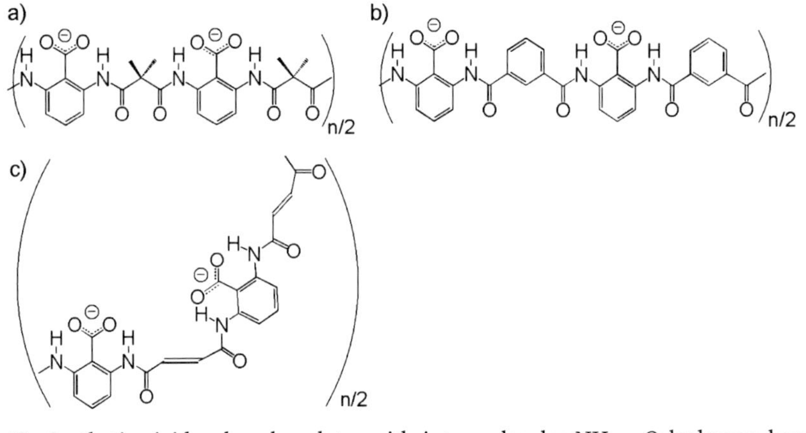

Fig. 13 Synthetic rigid polycarboxylates with intramolecular NH···O hydrogen bonds. **a** $\{NHC_6H_3(COO^-)NHCOC(CH_3)_2CO\}_n)$, **b** $\{NHC_6H_3(COO^-)NHCO\text{-}m\text{-}C_6H_4CO\}_n$, and **c** $\{NHC_6H_3(COO^-)NHCOCH=CHCO\}_n$

two ligands have carboxylate-parallel oriented moieties in the molecular dynamics minimized structures. In the moieties, the distances between adjacent carboxylates are almost 10 Å for $\{NHC_6H_3(COOH)NHCOC(CH_3)_2CO\}_n$ and 9 Å for $\{NHC_6H_3(COOH)NHCO\text{-}m\text{-}C_6H_4CO\}_n$, which are almost twice the distances of calcium ion separation in calcium carbonate. In contrast, $\{NHC_6H_3(COOH)NHCOCH=CHCO\}_n$ cannot possess a parallel-oriented carboxylate group because of the *trans* geometry of the fumaryl spacer. These results demonstrate the difficulty involved for $\{NHC_6H_3(COOH)NHCOCH=CHCO\}_n$ to interact with calcium carbonate at more than two points. Thus, the parallel-oriented carboxylate groups are crucial for making inorganic-organic composite.

11
Location of Strongly Binding Polycarboxylate on the Surface of a CaCO$_3$ Crystal

The CaCO$_3$/poly(1-carboxylate-3-*N*-*t*-BuNHCOC$_4$H$_6$) composite (Fig. 14a) was compared with that of poly(acrylate) (Fig. 14b). The intramolecularly hydrogen-bonded polymer gives a carboxylate CO ^{13}C-NMR signal at 186 ppm and an amide CO signal at 178 ppm, accompanied by a relatively weak CO$_3^{2-}$ signal at 169 ppm (Fig. 15). The remarkable loss of intensity of the CO$_3^{2-}$ ^{13}C signal is ascribed to sufficient cross-polarization [128]. The ^{13}C CP/MAS spectra of crystalline CaCO$_3$/poly(1-carboxylate-3-*N*-*n*-BuNHCOC$_4$H$_6$) were similar to the ^{13}C spectra of poly(1-carboxylate-3-*N*-*n*-BuNHCOC$_4$H$_6$ sodium salt) and poly(1-carboxyl-3-*N*-*t*-BuNHCOC$_4$H$_6$) in the solid state at 303 K. The composite shows a carboxylate CO signal at

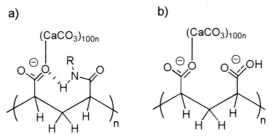

Fig. 14 a Amidated poly(acrylate)/CaCO₃ composite with 8-membered ring NH⋯O hydrogen bonds and **b** poly(acrylate)/CaCO₃ composite with the hydrogen bonds

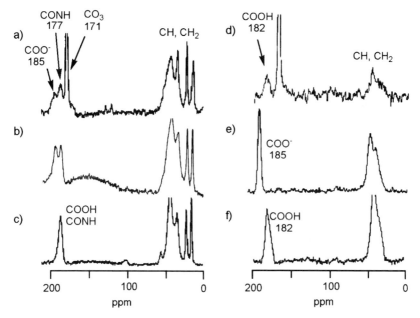

Fig. 15 Solid-state ^{13}C CP/MAS spectra of **a** CaCO₃/poly(1-carboxylate-3-N-n-BuNHCO C₄H₆) composite, **b** poly(1-carboxylate-3-N-n-BuNHCOC₄H₆ Na salt), **c** poly(1-carboxyl-3-N-t-BuNHCOC₄H₆), **d** CaCO₃/poly(acrylate) composite, **e** poly(acrylate Na salt), and **f** poly(acrylic acid) at 303 K

185 ppm and an amide CO signal at 178 ppm, indicating that the polymer ligand is bound to the crystals.

In comparison, the ^{13}C CP/MAS spectra of crystalline CaCO₃ containing poly(acrylate) do not show a clear carboxylate CO signal but rather a broad signal at 182 ∼ 184 ppm. A clear carboxylate COO⁻ signal for poly(acrylate sodium salt) is observed at 186 ppm and a carboxylic acid signal appears at 182 ppm in poly(acrylic acid) in the solid state at 303 K. The broad signal indicates the presence of a mixture of COO⁻ and COOH in the polymer chain. Thus, in the case of poly(acrylate), a part of the polymer ligand separates from the crystalline $CaCO_3$ to give a partially free poly(acid). These data are consistent with the reported finding that poly(acrylate) is involved in the nucleation of $CaCO_3$ crystallization but is then dislodged from the crystals [117]. The surface of crystalline $CaCO_3$ is thought to be covered with coordinated water molecules. The presence of their hydrogen atoms on the surface provides a hydrophobic environment around the Ca(II) ion [129]. Figure 16a shows a reported structure of the surface of a $CaCO_3$/stearate monolayer. When the pK_a of the carboxylic acid increases, the carboxylate readily converts to carboxylic acid and becomes dislodged from the crystals. This dislodgement is caused by hydrolysis of the Ca – O bond with water molecules and/or protons (Fig. 16b).

Poly[1-carboxylate-2-(N-t-butylcarbamoyl)ethylene-alt-ethylene] has a strong intramolecular 7-membered NH···O hydrogen bond in *threo* form (Fig. 17a) which is derived by the *cis*-opening of the maleic anhydride part in the precursor polymer [130]. A model compound, *threo*-(R,S)-2-(N-t-butylcarbamoyl)cyclohexanecarboxylic acid in *trans*-zigzag form similarly possesses a strong NH···O hydrogen bond (Fig. 17b). The polymer ligand provides vaterite crystals, and is not dislodged from the crystals even by washing with distilled water and methanol. The location of the Au-conjugated polymer ligand on the surface of, or inside the crystals of, novel gold colloid-containing poly(*p*-MeSC$_6$H$_4$NHCO-malate-*alt*-ethylene) having

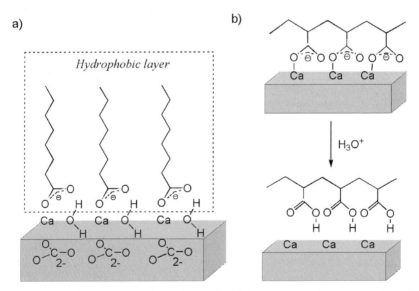

Fig. 16 a Reported structure for the surface of $CaCO_3$/stearate monolayer composite [129]. **b** Dislodgement of the polymer ligand from the surface of $CaCO_3$/poly(acrylate) composite with water washing under neutral conditions

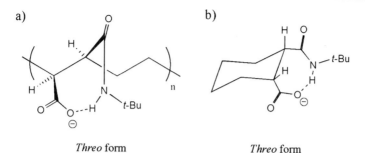

Fig. 17 a A possible structure of poly[1-carboxylate-2-(*N*-*t*-butylcarbamoyl)ethylene-*alt*-ethylene] in *threo* form and **b** a possible structure of (*S*,*R*)-2-(*N*-*t*-butylcarbamoyl)cyclohexanecarboxylate in *threo* form

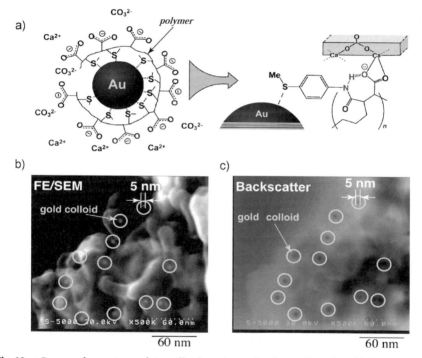

Fig. 18 a Proposed structure of Au colloid-conjugated polymer ligand and **b** the FE/SEM image of novel Au colloid-containing poly(*p*-MeSC$_6$H$_4$NHCOmalate-*alt*-ethylene)/CaCO$_3$ composite and its backscattering image (30 kV acceleration voltage, ×500 000). Au colloids are observed inside of *white circles*

an intramolecular 7-membered NH···O hydrogen bond was examined and a schematic structure of this composite is proposed in Fig. 18a. The composite consists of an agglomerate of nanosized vaterite crystals giving a FE/SEM and its backscatter (Fig. 18b,c). The data on the location of the Au-colloids in-

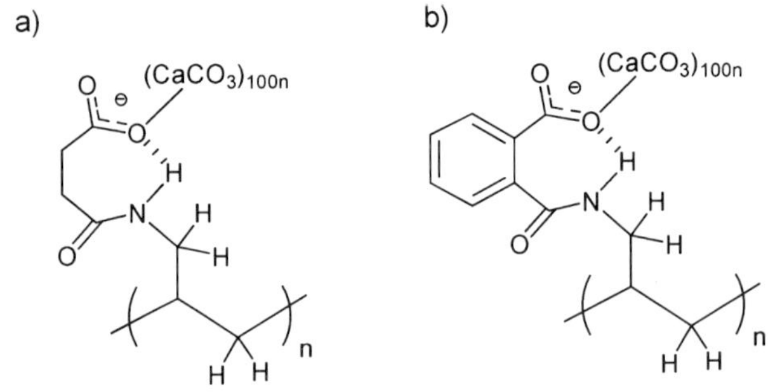

Fig. 19 Polycarboxylate/CaCO$_3$ composite with 7-membered ring NH···O hydrogen bonds including **a** succinate and **b** phthalate in side chains

dicate that the polymer ligand exists at the surface of fine vaterite crystals and oriented carboxylate ligands control the CaCO$_3$ polymorph (Takahashi et al., personal communication).

Poly(carboxylate) ligands using a poly(allylamine) skeleton, $\{CH_2CH(CH_2NHCO\text{-}R\text{-}COO^-)\}_n$ (R = CH$_2$CH$_2$, 2-Ph), have a strong 7-membered intramolecular NH···O hydrogen bond that also helps to stabilize the binding of the polymer ligand to vaterite CaCO$_3$ crystals (Fig. 19a,b) [131]. In comparison, the malonate polymer (R = CH=CH) produces calcite, which readily dislodges the polymer ligand from the edge of the calcite crystal with water/methanol washing. A structural change between the amide and phenyl planes converts from the NH···O hydrogen-bonded form into a higher basicity form that is readily hydrolyzed with water because of its high pK_a. The reason will be discussed in detail later.

12
Dependence of the CaCO$_3$ Composite on Polymer Ligand Tacticity

Various hydrogen-bond ring sizes were tested with 5-, 6- and 7-membered hydrogen-bonded polymer ligands that are derived from poly(methylmethacrylate). The poly[3-(2,2,2-triphenylacetylamino)propionate], poly[(Val-O$^-$)-MA], ligand with a weak intramolecular 5-membered hydrogen bond in the polymer side chain produces a calcite.

The syndiotactic-rich polymer, poly[3-(2,2,2-triphenylacetylamino)propionic acid] (*syndio*-poly[(β-Ala-OH)-MA]), and the isotactic-rich polymer, (*iso*-poly[(β-Ala-OH)-MA]), were examined to investigate the effect of polymer tacticity involving the side chain of the ligand upon CaCO$_3$ crystallization (Fig. 20). The diad contents for *syndio*-poly[(β-Ala-OH)-MA and *iso*-poly[(β-Ala-OH)-MA] were determined as 77% and 70%, respectively, determined by the ^{13}C NMR spectra in solution of the ester-formed polymers. An extremely broad amide NH ^{1}H-NMR signal for deprotonated *syndio*-poly[(β-Ala-O⁻)-MA] indicates the existence of intermolecular NH···O hydrogen bonds (Fig. 20a), whereas the large ¹H NMR shift of amide NH for deprotonated *iso*-poly[(β-Ala-O⁻)-MA] predicts the content of the stereoregular intramolecular NH···O hydrogen bonds without

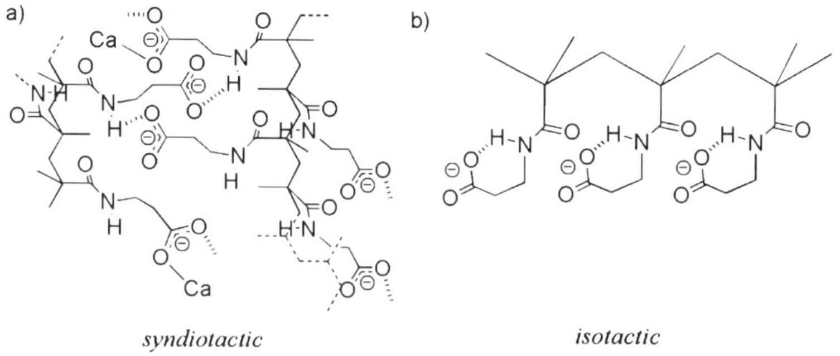

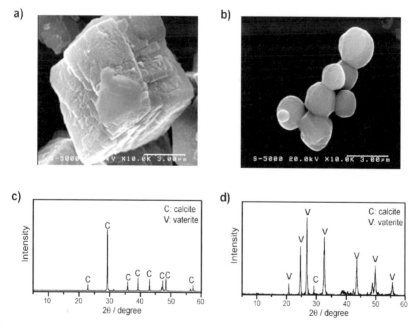

Fig. 20 Proposed solution structures of **a** *syndio*-poly[3-(methacryloylamino)propionate] anion with intermolecular NH···O hydrogen bonds and **b** *iso*-poly[3-(methacryloylamino)propionate] anion with intramolecular NH···O hydrogen bonds

Fig. 21 SEM images of **a** CaCO$_3$/*syndio*-poly[3-(methacryloylamino)propionate] composite (calcite) and **b** CaCO$_3$/*iso*-poly[3-(methacryloylamino)propionate] composite (vaterite). The XRD patterns of **c** CaCO$_3$/*syndio*-poly[3-(methacryloylamino)propionate] composite (calcite) and **d** CaCO$_3$/*iso*-poly[3-(methacryloylamino)propionate] composite (vaterite)

charge repulsion, when it is assumed that the MA main chain predominantly adopts a *trans* zigzag structure (Fig. 20b). A bulky model ligand, 3-(2,2,2-triphenylacetylamino)propionic acid, exhibits an amide NH signal at 6.31 ppm in chloroform-*d* at room temperature. Because of the formation of a strong NH $\cdots$ O hydrogen bond, the deprotonated anion form gives a shifted NH signal at 7.07 ppm. That is established by the observation of the hydrogen bond in the X-ray crystal structure of [NMe$_4$][3-(2,2,2-triphenylacetylamino)propionate]. In contrast, *syndio*-poly[(β-Ala-O$^-$)-MA] produces calcite and *iso*-poly[(β-Ala-O$^-$)-MA] serves vaterite as determined by the SEM images and X-ray diffraction (XRD) analysis (Fig. 21). Therefore, the *isotactic*-rich polymer is able to produce nanosized vaterite crystals because it possesses strong intramolecular NH $\cdots$ O hydrogen bonds.

Thus, the intramolecular NH $\cdots$ O hydrogen bond protects the Ca – O bond by shifting the polymer ligand pK_a and the partially covalent character against water. In addition to its strong Ca – O bonding, the orientation of the carboxylate groups along the edge of the CaCO$_3$ crystal determines the polymorph.

13
Biological Relevance of Synthetic Polymer Ligands

The binding properties of the synthetic polycarboxylate ligand have been demonstrated for Ca(II) complexes and biomineral model composites in terms of the intramolecular NH $\cdots$ O hydrogen bond. This hydrogen bond contributes not only to the increase of the formation constant by the lower shift of pK_a but also to the increase of Ca – O covalency upon neutral complexation. Both factors are supported in hydrophobic environments and even in the solid state, as analytical data have demonstrated for the CaCO$_3$/polymer composite in a low dielectric constant solvent. In nature, the dislodgement of a polymer ligand from the surface of biominerals often occurs with the coexistence of water. Therefore, both factors are crucial for preventing the Ca – O bond from hydrolysis with water.

In biominerals, there are many amide groups around the Ca-binding Glu or Asp carboxylate in biopolymer ligands. The biopolymer ligands presumably form intramolecular NH $\cdots$ O hydrogen bonds between the carboxylate oxyanion and the neighboring amide NH to lower the pK_a. As described above, lowering of the pK_a value increases the formation constant of the Ca – O bond in simple Ca complexes and on the surface of Ca biominerals. Fig. 22 illustrates a proposed structure around a Ca – O bond in the carboxylate ligand of a CaCO$_3$ biomineral.

In proteins, the hydrogen bond, e.g. NH $\cdots$ O $=$ C, is one of the most important parameters for controlling the shape of the proteins through the

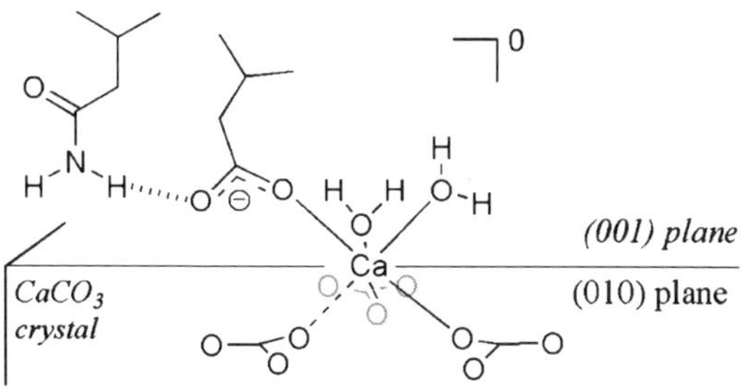

Fig. 22 Proposed structure on coordination of peptide ligand on the surface of CaCO$_3$ biomineral

hydrogen bond networks. The peptide conformation of Ca-binding fragments containing either Glu- or Asp produces a hairpin-like form that can readily convert to other structures to break the intramolecular NH···O hydrogen bond. The release and capture of the Ca ion are presumably associated with this regulation in peptide conformation as will be mentioned later.

14
Nanocomposite of the Nacreous Layer in *Pinctada Fucata*

Our data, which are based on simple Ca complexes and CaCO$_3$/synthetic polymer composites, indicate that nanosized vaterite crystals are formed mainly by strong Ca-binding of polymer ligands in addition to the matrix of ligand groups. In pearl, the nacreous layer consists of bricks. One large brick is believed to be an aragonite crystal because of its translucent appearance and diffraction analyses. Using high-resolution transmission electron microscopy (TEM), aragonite crystals from the nacreous layer in *Pinctada fucata* were examined [132]. Data for a thinly sliced nacreous layer containing a small brick (10 × 6 × 0.5 μm) were collected without force damage. The FE/TEM image of the pearl nacreous layer shows a clear boundary layer between two bricks (Fig. 23a). This boundary layer is known to be composed of biopolymers. However, most bricks contain many nanosized aragonite crystals, although large aragonite crystals have also been observed. Fig. 23b,c show elemental mapping images for Ca and N obtained [127] by electron energy loss spectroscopy (EELS). Ca spots that are 10 to 15 nm in size are caused by CaCO$_3$ crystals, whereas N spots that are about the same size probably come from amino acid residues of biopolymers in the range of 10 to 15 nm. Thus, the brick is not a single crystal but contains many nanosized aragonite crystals, similar to nanosized vaterite produced from synthetic polycarboxylate with intramolecular NH···O hydrogen bonds. In addition, the solid-state ^{13}C NMR spectrum of the nacreous layer suggests the presence of amino acid residues as a biopolymer. These biopolymers are lo-

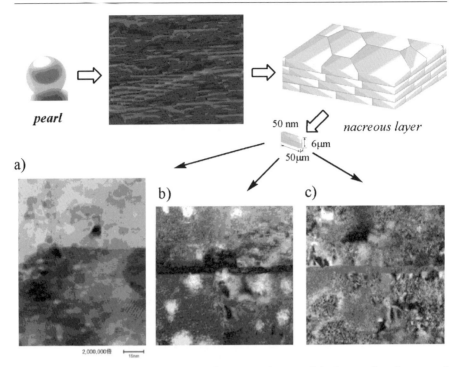

Fig. 23 a High resolution TEM image of nacreous layer and **b** Ca core-loss image and **c** N core-loss image for the elemental mapping of nacreous layer by EELS

cated in the boundary layer between two bricks and even inside the bricks as a nanosized ligand.

15
Variation of Metal–O Bonding by Conformational Change of Carboxylate Ligands

To elucidate the regulatory mechanism for mineralization involving biopolymers, the correlation between the function and the conformation of the peptide chains was studied. The formation and dissociation of Ca – O bonds seem to be controlled by the EF-hand conformation of Ca-binding proteins [56]. For example, a part of these structures has already been described in terms of the intramolecular NH $\cdots$ O hydrogen bond (Fig. 3).

In comparison, statherin is a multi-functional protein, which inhibits the primary and secondary precipitation of hydroxyapatite in saliva, as well as serving as a boundary lubricant [37]. On the basis of the NMR $T_1\rho$ measurements of the N-terminal 15-amino acid peptide in statherin, Drobny et al. proposed that the binding region in the *N*-terminus has rigid dynamics, which is associated with the dynamics of the following α-helix region [41]. However, such a conformational change of biopolymers to regulate mineralization is still unclear because there is no detailed information on the Ca-mineralizing protein.

16
Proton-Driven Conformation Switch in Asp-Oligopeptide and Model Compounds

The carboxylate anion functions not only as a coordinating ligand but also as a nucleophile in the active center of proteases (e.g., pepsin, which contains a Phe-Asp-Thr-Gly-Ser-Ser fragment, as well as an invariant amino acid fragment, Asp-Thr-Gly in HIV-1 proteases) [133, 134]. A crystallographic analysis of these proteins indicates that the invariant Asp-containing peptide fragments of Asp-X-Gly (X = Thr, Ser) form a hairpin turn with NH$\cdots$O hydrogen bonds. A model Asp-containing peptide, benzyloxycarbonyl-Phe-Asp(COO$^-$)-Thr-Gly-Ser-Ala-NHCy (Cy = cyclohexyl) anion, gives a hairpin turn structure in acetonitrile (Fig. 24). An Asp-containing tripeptide, AdCO-Asp(COOH)-Val-Gly-NHCH$_2$Ph (Ad = adamantly), has an inverse γ-turn structure in the aspartic acid state, whereas the tripeptide converts to a β-turn-like conformer with intramolecular NH$\cdots$O hydrogen bonds in chloroform or aqueous micellar solution (Fig. 24) [135]. This type of conformational change is not detected between AdCO-Asp(COOH)-Val-NHAd and its anion state, AdCO-Asp(COO$^-$)-Val-NHAd. The model peptide study suggests that the hydrogen bond stabilizes the anion state to decrease the basicity of the carboxylate anion, presumably resulting in a decrease of nucleophilicity.

It is difficult to elucidate the energy barrier in a proton-driven conformational change from the carboxylic acid structure to the carboxylate anion form in proteins and even in oligopeptides. As a simple model compound, the doubly amidated Kemp's acid derivative, r-1,c-3,c-5-(CH$_3$)$_3$-3,5-(Ph$_2$CHNHCO)$_2$C$_6$H$_6$-1-COOH, was synthesized [136]. A chair form of the protonated compound converts to a twist-boat form with deprotonation as shown in Fig. 25. The energy barrier for the conformational transformation after deprotonation is approximately 40 ∼ 80 kJ/mol in acetonitrile at room temperature. Another proton-driven conformational change is observed between a monoamidated maleic acid and its maleate. After deprotonation, conformer A is energetically unstable because the carboxylate has a higher basicity of oxyanion with the negative charge of the neighboring carbonyl oxygen. Then, as shown in Fig. 25, the change occurs by the rotation between the amide plane and the C = C plane over a relatively large energy barrier. Conformer A has a long lifetime when compared with that of monoamidated succinate.

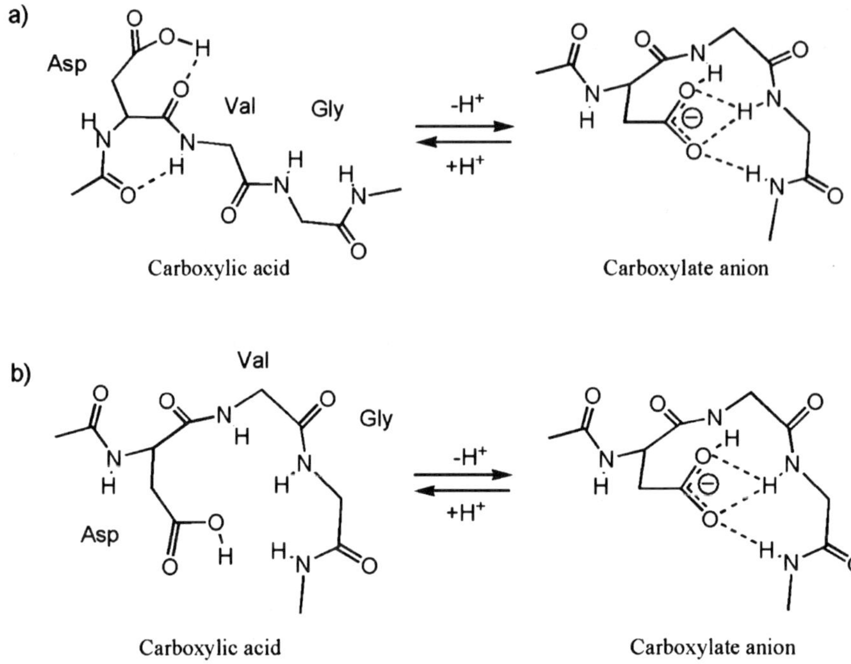

Fig. 24 Proton-driven conformational changes of AdCO-Asp(COOH)-Val-Gly-NHCH$_2$Ph and its anion form **a** in chloroform and **b** in aqueous micellar solution

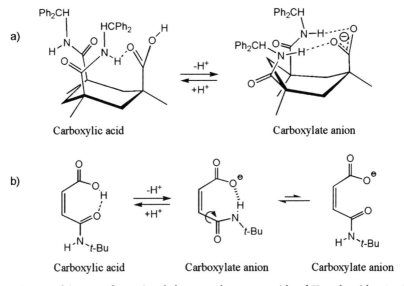

Fig. 25 Proton-driven conformational changes **a** between amidated Kemp's acid, r-1,c-3,c-5-(CH$_3$)$_3$-3,5-(Ph$_2$CHNHCO)$_2$C$_6$H$_6$-1-COOH, and its anion form, and **b** between (Z)-4-(t-butylamino)-4-oxo-2-butenoic acid and its anion form

17
Binding Regulation of Polycarboxylate on CaCO$_3$ Crystals by Conformational Change

Maleic acid containing polycarboxylate, poly[(Z)-3-allylcarbamoyl-2-propionic acid], produces the CaCO$_3$/polycarboxylate composite (calcite) at pH 3.8 ∼ 4.0 [137]. The composite readily dislodges the polymer ligand with water-washing, although the maleate part of the polycarboxylate has a strong intramolecular NH···O hydrogen bond. The SEM images in Fig. 26a and b show that a large calcite crystal having many swells on the edge and surface before washed with water. A ^{13}C NMR spectrum shows that the composite contains the polymer ligand before washing, whereas the polymer ligand is dislodged from the surface creating a hollow after washing is complete (Fig. 26b). The hollow has a 50 nm diameter and contains 20 polymer chains when it is assumed that one polymer chain adopts a *trans*-zigzag structure with a 3.5-nm diameter sphere. Crystal nucleation starts from an agglomerate of polycarboxylate, and the crystal growth proceeds on the opposite side from the polycarboxylate-binding site. However, weak interaction between the polymer ligand and the crystal surface leads to the formation of calcite and easily dislodges the polymer ligand from the surface.

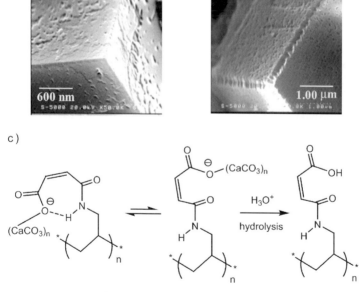

Fig. 26 FE/SEM images of CaCO$_3$/poly[(Z)-3-allylcarbamoyl-2-propenate] composite **a** before washing with MeOH/water and **b** after the washing. **c** The schematic conformational change of poly[(Z)-3-allylcarbamoyl-2-propenate] on the surface of CaCO$_3$ composite

As described previously, the maleate anion produces a strong intramolecular NH $\cdots$ O hydrogen bond between the carboxylate anion and the amide NH. The hydrogen bond supports a lower pK_a shift to prevent the Ca–O bond from dissociating. Actually, the succinate-containing polycarboxylate/ $CaCO_3$ composite does not dislodge the ligand even with washing under neutral conditions. The easy dislodgement of the ligand is caused by a conformational change from a stable form having a strong intramolecular NH $\cdots$ O hydrogen bond. The maleate ligand possesses an equilibrium between conformer A and conformer B (Fig. 26c). The ligand of conformer B has a higher basicity, which leads to easy hydrolysis of the Ca–O bond. The lifetime of conformer B is considered to be long enough for hydrolysis to occur. The weak interaction caused by coexistence of conformer B affects crystal growth to form calcite and the easy dislodgement of the polymer ligand with washing. These results support that a strong Ca composite requires a strong intramolecular NH $\cdots$ O hydrogen bond, especially in the presence of prelocated amide NH for lowering pK_a. In addition, it is required that no conformer containing carboxylate with a high basicity is present.

In contrast, when the polymer ligand has a conformer with high basicity and a long lifetime as well as a hydrogen-bonded stable conformer, the $CaCO_3$ composite is unstable. This instability is associated with the fact that native Ca biominerals often dissociate Ca ions by a conformational change of proteins in a biological system. Presumably, the strong intramolecular NH $\cdots$ O hydrogen bond plays an important role in Ca regulation.

18
Conclusions

A carboxylate anion without any intramolecular NH $\cdots$ O hydrogen bond has a high basicity under hydrophobic conditions when the corresponding carboxylic acid has a high pK_a value. The residual negative charge on the carboxylate oxyanion of the Ca–O bond leads to the formation of a polymeric cluster. The high basicity is readily decreased by the intramolecular NH $\cdots$ O hydrogen bonds. The prelocated amide NH decreases the pK_a value of acid. The decrease of the pK_a value results in an increase of the formation constant in a Ca-carboxylate complexation. Then the hydrogen bond controls coordination/dissociation of metal–oxygen bonds. In the case of neutral complexation for the Ca-carboxylate complexes, the NH $\cdots$ O hydrogen bond supports the covalent character of the Ca–O bond. Both effects contribute to the strong Ca–O binding on the surface of $CaCO_3$/polycarboxylate with intramolecular NH $\cdots$ O hydrogen bonds.

Ca–O bonds can be regulated by the NH $\cdots$ O hydrogen bonds and the successive hydrogen bonding networks in peptide chains. The combined results in carboxylate, sulfonate and phosphate Ca complexes suggest the presence of a communication between the basicity of the oxyanion in carboxylate and hydrogen bonds cooperating with the peptide conformation. In addition, the controlled binding of metal ions by the NH $\cdots$ O hydrogen bonds contributes to the crystallization of inorganic composites. Therefore, the hydrogen bonds organized by peptide conformations seem to be an essential factor for the regulatory formation of nano-architecture and these observations help in elucidating the interface structure of inorganic and organic phases in biominerals.

References

1. Frankel RB, Mann S (1994) In: King RB (ed) Encyclopedia of Inorganic Chemistry. Wiley, New York, p 269
2. Wada K, Kobayashi I (1996) Biomineralization and Hard Tissue of Marine Organisms. Tokaidaigaku-shuppankai, Tokyo
3. Berman A, Hanson J, Leiserowitz L, Koetzle TF, Weiner S, Addadi L (1993) J Phys Chem 97:5162
4. Aizenberg J, Albeck S, Weiner S, Addadi L (1994) J Cryst Growth 142:156
5. Addadi L, Weiner S (1985) Proc Natl Acad Sci USA 82:4110
6. Albeck S, Aizenberg J, Addadi L, Weiner S (1993) J Am Chem Soc 115:11691
7. Berman A, Addadi L, Weiner S (1988) Nature 331:546
8. Lowenstam HA (1981) Science 211:1126
9. Mann S (1988) Nature 332:119
10. Lowenstam H, Weiner S (1989) On Biomineralization. Oxford University Press, New York
11. Wada K (1972) Biomineralization 6:141
12. Miyamoto H, Miyashita T, Okushima M, Nakano S, Morita T, Matushiro A (1996) Proc Natl Acad Sci USA 93:9657
13. Mann S (1989) In: Mann S, Webb J, Williams RJP (eds) Biomineralization. Wiley, New York
14. De Jong E, Bosch L, Westbroek JP (1976) Eur J Biochem 70:611
15. Marsh ME, Chang D, King GC (1992) J Biol Chem 267:20507
16. Traub W, Arad T, Weiner S (1989) Proc Natl Acad Sci USA 86:9822
17. Sasaki N, Sudoh Y (1997) Calif Tissue Int 60:361
18. Robey PG (1997) Connect Tissue Res 35:131
19. Butler WT, Ritche HH, Bronckers ALJJ (1997) Dental Enamel (Ciba Foundation Symposium 205) 205:107
20. Fueredi-Milhofer H, Moradian-Oldak J, Weiner S, Veis A, Minits KP, Addadi L (1994) Connect Tissue Res 30:251
21. George A, Bannon L, Sabsay B, Dillon JW, Malone J, Veis A, Jenkins NA, Gilbert DJ, Copeland NG (1996) J Biol Chem 271:32869
22. Addadi L, Weiner S (1992) Angew Chem Int Ed Engl 31:153
23. Mann S, Perry CC (1991) Adv Inorg Chem 36:137
24. Brecevie L, Nothig-Laslo V, Kralj D, Popovic S (1996) J Chem Soc Faraday Trans 92:1017
25. Thompson JB, Paloczi GT, Kindt JH, Michenfelder M, Smith BL, Stucky G, Morse DE, Hansma PK (2000) Biophys J 79:3307

26. Naka K, Chujo Y (2001) Chem Mater 13:3245
27. Sinha A, Agrawal A, Das SK, Ravikumar B, Rao V, Ramachanndrarao P (2001) J Mater Sci Lett 20:1569
28. Naka K, Chujo Y (2003) C R Chimie 6:1193
29. Kim W, Robertson RE, Zand R (2005) Cryst Growth Des 5:513
30. Han JT, Xu X, Kim DH, Cho K (2005) Chem Mater 17:136
31. Wei H, Shen Q, Zhao Y, Zhou Y, Wang D, Xu D (2004) J Cryst Growth 264:424
32. Ishii K, Tsutsui N, Watanabe T, Yanagisawa T, Nagasawa H (1998) Biosci Biotechnol Biochem 62:291
33. Li C, Bostsaris GD, Kaplan DL (2002) Cryst Growth Des 2:387
34. Ishii K, Yanagisawa T, Nagasawa H (1996) Biosci Biotechnol Biochem 60:1479
35. Levi-Kalisman Y, Raz S, Weiner S, Addadi L, Sagi I (2002) Adv Funct Mater 12:43
36. Aizenberg J, Lambert G, Weiner S, Addadi L (2002) J Am Chem Soc 124:32
37. Raj PA, Johnsson M, Lemine MJ, Nancollas GH (1992) J Biol Chem 267:5968
38. Schlesinger DH, Hay DI, Lemine MJ (1989) Int J Pept Protein Res 34:374
39. Long JR, Dindot JL, Zebroski H, Kiihne S, Clark RH, Campbell AA, Stayton PS, Drobny GP (1998) Proc Natl Acad Sci USA 95:12083
40. Long JR, Shaw WJ, Stayton PS, Drobny GP (2001) Biochemistry 40:15451
41. Shaw WJ, Long JR, Campbell AA, Stayton PS, Drobny GP (2000) J Am Chem Soc 122:7118
42. Okamura T, Ueyama N, Nakamura A, Ainscough EW, Brodie AM, Waters JM (1993) J Chem Soc Chem Commun, p 1685
43. Ueyama N, Okamura T, Nakamura A (1992) J Am Chem Soc 114:8129
44. Ueyama N, Okamura T, Nakamura A (1992) Chem Commun, p 1019
45. Okamura T, Takamizawa S, Ueyama N, Nakamura A (1998) Inorg Chem 37:18
46. Ueyama N, Nishikawa N, Yamada Y, Okamura T, Nakamura A (1996) J Am Chem Soc 118:12826
47. Okamura T, Ueyama N, Nakamura A (1998) Chem Lett, p 199
48. Dey A, Okamura T, Ueyama N, Hedman B, Hodgson KO, Solomon EI (2005) J Am Chem Soc 127:12046
49. Ueyama N, Terakawa T, Nakata M, Nakamura A (1983) J Am Chem Soc 105:7098
50. Ueyama N, Nakata M, Fuji M, Terakawa T, Nakamura A (1985) Inorg Chem 24:2190
51. Ueno T, Kousumi Y, Yoshizawa-Kumagaye K, Nakajima K, Ueyama N, Okamura T, Nakamura A (1998) J Am Chem Soc 120:12264
52. Ueyama N, Nishikawa N, Yamada Y, Okamura T, Nakamura A (1998) Inorg Chim Acta 283:91
53. Murthy MRN, Reid-III TJ, Sicignano A, Tanaka N, Rossmann MG (1981) J Mol Biol 152:465
54. Fita I, Rossmann MG (1985) J Mol Biol 185:21
55. Kanamori D, Yamada Y, Okamura T, Yamamoto H (2003) Inorg Chim Acta 358:331
56. Strynadka NCJ, James MNG (1994) In: King RB (ed) Encyclopedia of Inorganic Chemistry. Wiley, New York
57. Essen L, Perisic O, Katan M, Wu Y, Roberts MF, Williams RL (1997) Biochemistry 36:1704
58. Weston SA, Lahm A, Suck D (1992) J Mol Biol 226:1237
59. Oefner C, Suck D (1986) J Mol Biol 192:605
60. Copley RR, Barton GJ (1994) J Mol Biol 242:321
61. Pflugrath JW, Quiocho FA (1988) J Mol Biol 200:163

62. Jacobson BL, Quiocho FA (1988) J Mol Biol 204:783
63. Kanamori D, Furukawa A, Okamura T, Yamamoto H, Ueyama N (2005) Org Biomol Chem 3:1453
64. Onoda A, Yamada Y, Takeda J, Nakayama Y, Okamura T, Doi M, Ueyama N (2004) Bull Chem Soc Japan 77:321
65. Ueyama N, Takeda J, Yamada Y, Onoda A, Okamura T, Nakamura A (1999) Inorg Chem 38:475
66. Ueyama N, Hosoi T, Yamada Y, Doi M, Okamura T, Nakamura A (1998) Macromolecules 31:7119
67. Ueyama N, Yamada Y, Takeda J, Okamura T, Mori W, Namakura A (1996) J Chem Soc, Chem Commun, p 1377
68. Ueyama N, Nishikawa N, Yamada Y, Okamura T, Nakamura A (1996) J Am Chem Soc 118:12826
69. Onoda A, Yamada Y, Takeda J, Nakayama Y, Okamura T, Doi M, Yamamoto H, Ueyama N (2004) Bull Chem Soc Jpn 77:321
70. Onoda A, Yamada Y, Nakayama Y, Takahashi K, Okamura T, Nakamura A, Yamamoto H, Ueyama N (2004) Inorg Chem 43:4447
71. Chatterjee A, Maslen EN, Watoson KJ (1988) Acta Crystallogr, Sect B: Struct Sci 44:386
72. Chatterjee A, Maslen EN, Watoson KJ (1988) Acta Crystallogr, Sect B: Struct Sci 44:381
73. Onoda A, Yamada Y, Doi M, Okamura T, Ueyama N (2001) Inorg Chem, p 516
74. Onoda A, Yamada Y, Okamura T, Doi M, Yamamoto H, Ueyama N (2001) J Am Chem Soc 124:1052
75. Einspahr H, Bugg CE (1981) Acta Crystallogr B 37:1044
76. Balasubramanian T, Muthuah PT (1996) Acta Crystallogr Sect C 52:2072
77. Russell VA, Etter MC, Ward MD (1994) J Am Chem Soc 116:1941
78. Russell VA, Ward MD (1996) Acta Crystallogr Sect B 52:209
79. Russell VA, Evans CC, Li W, Ward MD (1997) Science 276:575
80. Ojala WH, Lu LK, Albers KE, Gleason WB (1994) Acta Crystallogr Sect B 50:684
81. Kelly TR, Kim MH (1994) J Am Chem Soc 116:7072
82. Valiyaveettil S, Engbersen JFJ, Verboom W, Reinhoudt DN (1993) Angew Chem Int Ed Engl 32:900
83. Alebertsson J, Osakarsson A, Svensson C (1978) Acta Cryst B 34:2737
84. Whitlow S (1972) Acta Cryst B 28:1914
85. Uchtman VA, Jandacek R (1980) Inorg Chem 19:350
86. Taga T, Osaki K (1976) Bull Chem Soc Japan 49:1517
87. Schuckmann W, Fuess H, Bats JW (1978) Acta Cryst B 34:3754
88. Oonishi I, Sato S, Saito Y (1975) Acta Cryst B 31:1318
89. Matsuzaki T, Iitaka Y (1971) Acta Cryst B 28:1977
90. Gputa MP, Prasad N (1976) Acta Cryst B 32:3257
91. Flapper WMJ, Verschoor GC, Rutten EWM, Romers C (1977) Acta Cryst B 33:5
92. Einspahr H, Gartland GL, Bugg CE (1977) Acta Cryst B 33:3385
93. Barnett BL, Uchtman VA (1979) Inorg Chem 18:2674
94. Adair BA, De Delgado GD, Delgado JM, Cheetham AK (2000) Angew Chem Int Ed 39:745
95. Cao G, Lee H, Lynch VM, Mallouk TE (1988) Inorg Chem 27:2781
96. Cheetham AK, Fe'rey G, Loisear T (1999) Angew Chem Int Ed 38:3268
97. Muller A, Hovemeier K, Krickemeyer E, Bogge H (1995) Angew Chem Int Ed Engl 34:779

98. Natarajan S, Neeraj S, Choudhury A, Rao CNR (2000) Inorg Chem 39:1426
99. Neeraj S, Natarajan S, Rao CNR (2000) J Solid State Chem 150:417
100. Rao CNR, Natarajan S, Choudrury A, Neeraj S, Ayi AA (2001) Acc Chem Res 34:80
101. Walawalkar MG, Murugavel R, Roesky HW, Uson I, Kraetzner R (1998) Inorg Chem 473
102. Yan W, Yu J, Shi Z, Xu R (2000) Chem Commun, p 1431
103. Shubnell AJ, Kosnic EJ, Squattrito PJ (1994) Inorg Chim Acta 216:101
104. Kennedy AR, Mcnair C, Smith WE, Chisholm G, Teat SJ (2000) Angew Chem Int Ed 39:638
105. Shakeri V, Haussuhl S (1992) Z Kristallogr 198:299
106. Gunderman BJ, Squattrito PJ (1996) Acta Crystallogr Sect C
107. Oliver S, Kuperman A, Ozin GA (1998) Angew Chem Int Ed Engl 37:46
108. Rao CNR, Natarajan S, Neeraj S (2000) J Am Chem Soc 122:2810
109. Rao CNR, Natarajan S, Neeraj S (2000) J Solid State Chem 152:302
110. Bissinger P, Kumberger O, Schier A (1990) Chem Ber 124:509
111. Sato T (1984) Acta Crystallogr C 40:738
112. Cao G, Lynch VM, Swinnea S, Mallouk TE (1990) Inorg Chem 29:2112
113. Rudolf PR, Clarke ET, Martell AE, Clearfield A (1988) Acta Crystallogr C 44:796
114. Holmes MA, Matthews BM (1982) J Mol Biol 160:623
115. Shao X, Davletov BA, Sutton RB, Sudhof TC, Rizo J (1996) Science 273:248
116. Voordouw G, Roche RS (1974) Biochemistry 13:5017
117. Addadi L, Moradian-Oldak J, Shay E, Maroudas NG, Weiner S (1987) Proc Natl Acad Sci USA 84:2732
118. Albeck S, Weiner S, Addadi L (1987) Chem Eur J 2:278
119. Sugawara A, Kato T (2000) Chem Commun, p 487
120. Xu G-F, Yao N, Aksay IA, Groves JT (1998) J Am Chem Soc 120:11977
121. Kuther J, Sechadri R, Nelles G, Assenmacher W, Butt H-J, Mader W, Tremel W (1999) Chem Mater 11:1317
122. Donners JJJM, Nolte RJM, Sommerdijk NAJM (2002) J Am Chem Soc 124:9700
123. Falini G (2000) Int J Inorg Mater 2:455
124. Kato T, Sugawara A, Hosoda N (2002) Adv Mater 14:869
125. Naka K, Dong D-K, Tanaka Y, Chujo Y (2000) J Chem Soc Chem Commun, p 1537
126. DeOliveira DB, Laursen RA (1997) J Am Chem Soc 119:10627
127. Yamamoto H, Ueyama N (2004) In: Ueyama N, Harada A (eds) Nano-Structured Macromolecular Materials. Kodansha Ltd and Springer Inc, Tokyo
128. Ueyama N, Kozuki H, Takahashi K, Onoda A, Okamura T, Yamamoto H (2001) Macromolecules 34:2607
129. Wheeler AP, Rusenko KW, George JW, Sykes CS (1987) Comp Biochem Physiol 87:953
130. Takahashi K, Kozuki H, Onoda A, Okamura T, Yamamoto H, Ueyama N (2002) J Inorg Organomet Polym 12:99
131. Takahashi K, Doi M, Kobayashi A, Taguchi T, Onoda A, Okamura T, Yamamoto H, Ueyama N (2004) J Cryst Growth 263:552
132. Takahashi K, Yamamoto H, Onoda A, Doi M, Inaba T, Chiba M, Kobayashi A, Taguchi T, Okamura T, Ueyama N (2004) Chem Commun, p 996
133. Suguna K, Bott RR, Padlan EA, Subramanian E, Sheriff S, Cohen GH, Davies DR (1987) J Mol Biol 196:877
134. Wlodawer A, Miller M, Jasko'lski M, Sathyanarayana BK, Baldwin E, Weber IT, SL M, Clawson L, Schneider J, Kent SBH (1989) Science 245:616

135. Onoda A, Yamamoto H, Yamada Y, Lee K, Adachi S, Okamura T, Yoshizawa-Kumagaye K, Nakajima K, Kawakami T, Aimoto S, Ueyama N (2005) Biopolymers 80:233
136. Onoda A, Haruna H, Yamamoto H, Takahashi K, Kozuki H, Okamura T, Ueyama N (2005) Eur J Org Chem, p 641
137. Takahashi K, Doi M, Kobayashi A, Taguchi T, Onoda A, Okamura T, Yamamoto H, Ueyama N (2004) Chem Lett 33:192
138. Einspahr H, Bugg CE (1981) Acta Crystallogr, Sect B: Struct Sci 57:1044
139. Li C-T, Caughlan CN (1965) Acta Crystallogr 19:637

Author Index Volumes 251–271

Author Index Vols. 26–50 see Vol. 50
Author Index Vols. 51–100 see Vol. 100
Author Index Vols. 101–150 see Vol. 150
Author Index Vols. 151–200 see Vol. 200
Author Index Vols. 201–250 see Vol. 250

The volume numbers are printed in italics

Ajayaghosh A, George SJ, Schenning APHJ (2005) Hydrogen-Bonded Assemblies of Dyes and Extended π-Conjugated Systems. *258*: 83–118
Albert M, Fensterbank L, Lacôte E, Malacria M (2006) Tandem Radical Reactions. *264*: 1–62
Alberto R (2005) New Organometallic Technetium Complexes for Radiopharmaceutical Imaging. *252*: 1–44
Alegret S, see Pividori MI (2005) *260*: 1–36
Amabilino DB, Veciana J (2006) Supramolecular Chiral Functional Materials. *265*: 253–302
Anderson CJ, see Li WP (2005) *252*: 179–192
Anslyn EV, see Houk RJT (2005) *255*: 199–229
Appukkuttan P, Van der Eycken E (2006) Microwave-Assisted Natural Product Chemistry. *266*: 1–47
Araki K, Yoshikawa I (2005) Nucleobase-Containing Gelators. *256*: 133–165
Armitage BA (2005) Cyanine Dye–DNA Interactions: Intercalation, Groove Binding and Aggregation. *253*: 55–76
Arya DP (2005) Aminoglycoside–Nucleic Acid Interactions: The Case for Neomycin. *253*: 149–178

Bailly C, see Dias N (2005) *253*: 89–108
Balaban TS, Tamiaki H, Holzwarth AR (2005) Chlorins Programmed for Self-Assembly. *258*: 1–38
Balzani V, Credi A, Ferrer B, Silvi S, Venturi M (2005) Artificial Molecular Motors and Machines: Design Principles and Prototype Systems. *262*: 1–27
Barbieri CM, see Pilch DS (2005) *253*: 179–204
Barchuk A, see Daasbjerg K (2006) *263*: 39–70
Bayly SR, see Beer PD (2005) *255*: 125–162
Beer PD, Bayly SR (2005) Anion Sensing by Metal-Based Receptors. *255*: 125–162
Bertini L, Bruschi M, De Gioia L, Fantucci P, Greco C, Zampella G (2007) Quantum Chemical Investigations of Reaction Paths of Metalloenzymes and Biomimetic Models – The Hydrogenase Example. *268*
Bier FF, see Heise C (2005) *261*: 1–25
Blum LJ, see Marquette CA (2005) *261*: 113–129
Boiteau L, see Pascal R (2005) *259*: 69–122
Bolhuis PG, see Dellago C (2007) *268*
Borovkov VV, Inoue Y (2006) Supramolecular Chirogenesis in Host–Guest Systems Containing Porphyrinoids. *265*: 89–146

Boschi A, Duatti A, Uccelli L (2005) Development of Technetium-99m and Rhenium-188 Radiopharmaceuticals Containing a Terminal Metal–Nitrido Multiple Bond for Diagnosis and Therapy. *252*: 85–115

Braga D, D'Addario D, Giaffreda SL, Maini L, Polito M, Grepioni F (2005) Intra-Solid and Inter-Solid Reactions of Molecular Crystals: a Green Route to Crystal Engineering. *254*: 71–94

Brebion F, see Crich D (2006) *263*: 1–38

Brizard A, Oda R, Huc I (2005) Chirality Effects in Self-assembled Fibrillar Networks. *256*: 167–218

Bruce IJ, see del Campo A (2005) *260*: 77–111

Bruschi M, see Bertini L (2007) *268*

del Campo A, Bruce IJ (2005) Substrate Patterning and Activation Strategies for DNA Chip Fabrication. *260*: 77–111

Carney CK, Harry SR, Sewell SL, Wright DW (2007) Detoxification Biominerals. *270*: 155–185

Chaires JB (2005) Structural Selectivity of Drug-Nucleic Acid Interactions Probed by Competition Dialysis. *253*: 33–53

Chiorboli C, Indelli MT, Scandola F (2005) Photoinduced Electron/Energy Transfer Across Molecular Bridges in Binuclear Metal Complexes. *257*: 63–102

Cölfen H (2007) Bio-inspired Mineralization Using Hydrophilic Polymers. *271*: 1–77

Collin J-P, Heitz V, Sauvage J-P (2005) Transition-Metal-Complexed Catenanes and Rotaxanes in Motion: Towards Molecular Machines. *262*: 29–62

Collyer SD, see Davis F (2005) *255*: 97–124

Commeyras A, see Pascal R (2005) *259*: 69–122

Coquerel G (2007) Preferential Crystallization. *269*

Correia JDG, see Santos I (2005) *252*: 45–84

Costanzo G, see Saladino R (2005) *259*: 29–68

Credi A, see Balzani V (2005) *262*: 1–27

Crestini C, see Saladino R (2005) *259*: 29–68

Crich D, Brebion F, Suk D-H (2006) Generation of Alkene Radical Cations by Heterolysis of β-Substituted Radicals: Mechanism, Stereochemistry, and Applications in Synthesis. *263*: 1–38

Cuerva JM, Justicia J, Oller-López JL, Oltra JE (2006) Cp_2TiCl in Natural Product Synthesis. *264*: 63–92

Daasbjerg K, Svith H, Grimme S, Gerenkamp M, Mück-Lichtenfeld C, Gansäuer A, Barchuk A (2006) The Mechanism of Epoxide Opening through Electron Transfer: Experiment and Theory in Concert. *263*: 39–70

D'Addario D, see Braga D (2005) *254*: 71–94

Danishefsky SJ, see Warren JD (2007) *267*

Darmency V, Renaud P (2006) Tin-Free Radical Reactions Mediated by Organoboron Compounds. *263*: 71–106

Davis F, Collyer SD, Higson SPJ (2005) The Construction and Operation of Anion Sensors: Current Status and Future Perspectives. *255*: 97–124

Deamer DW, Dworkin JP (2005) Chemistry and Physics of Primitive Membranes. *259*: 1–27

Dellago C, Bolhuis PG (2007) Transition Path Sampling Simulations of Biological Systems. *268*

Deng J-Y, see Zhang X-E (2005) *261*: 169–190

Dervan PB, Poulin-Kerstien AT, Fechter EJ, Edelson BS (2005) Regulation of Gene Expression by Synthetic DNA-Binding Ligands. *253*: 1–31

Dias N, Vezin H, Lansiaux A, Bailly C (2005) Topoisomerase Inhibitors of Marine Origin and Their Potential Use as Anticancer Agents. *253*: 89–108
DiMauro E, see Saladino R (2005) *259*: 29–68
Dittrich M, Yu J, Schulten K (2007) PcrA Helicase, a Molecular Motor Studied from the Electronic to the Functional Level. *268*
Dobrawa R, see You C-C (2005) *258*: 39–82
Du Q, Larsson O, Swerdlow H, Liang Z (2005) DNA Immobilization: Silanized Nucleic Acids and Nanoprinting. *261*: 45–61
Duatti A, see Boschi A (2005) *252*: 85–115
Dworkin JP, see Deamer DW (2005) *259*: 1–27

Edelson BS, see Dervan PB (2005) *253*: 1–31
Edwards DS, see Liu S (2005) *252*: 193–216
Ernst K-H (2006) Supramolecular Surface Chirality. *265*: 209–252
Ersmark K, see Wannberg J (2006) *266*: 167–197
Escudé C, Sun J-S (2005) DNA Major Groove Binders: Triple Helix-Forming Oligonucleotides, Triple Helix-Specific DNA Ligands and Cleaving Agents. *253*: 109–148
Van der Eycken E, see Appukkuttan P (2006) *266*: 1–47

Fages F, Vögtle F, Žinić M (2005) Systematic Design of Amide- and Urea-Type Gelators with Tailored Properties. *256*: 77–131
Fages F, see Žinić M (2005) *256*: 39–76
Faigl F, Schindler J, Fogassy E (2007) Advantages of Structural Similarities of the Reactants in Optical Resolution Processes. *269*
Fantucci P, see Bertini L (2007) *268*
Fechter EJ, see Dervan PB (2005) *253*: 1–31
Fensterbank L, see Albert M (2006) *264*: 1–62
Fernández JM, see Moonen NNP (2005) *262*: 99–132
Fernando C, see Szathmáry E (2005) *259*: 167–211
Ferrer B, see Balzani V (2005) *262*: 1–27
De Feyter S, De Schryver F (2005) Two-Dimensional Dye Assemblies on Surfaces Studied by Scanning Tunneling Microscopy. *258*: 205–255
Flood AH, see Moonen NNP (2005) *262*: 99–132
Fogassy E, see (2007) *269*
Fricke M, Volkmer D (2007) Crystallization of Calcium Carbonate Beneath Insoluble Monolayers: Suitable Models of Mineral–Matrix Interactions in Biomineralization? *270*: 1–41
Fujimoto D, see Tamura R (2007) *269*
Fujiwara S-i, Kambe N (2005) Thio-, Seleno-, and Telluro-Carboxylic Acid Esters. *251*: 87–140

Gansäuer A, see Daasbjerg K (2006) *263*: 39–70
Garcia-Garibay MA, see Karlen SD (2005) *262*: 179–227
Gelinck GH, see Grozema FC (2005) *257*: 135–164
Geng X, see Warren JD (2007) *267*
George SJ, see Ajayaghosh A (2005) *258*: 83–118
Gerenkamp M, see Daasbjerg K (2006) *263*: 39–70
Giaffreda SL, see Braga D (2005) *254*: 71–94
De Gioia L, see Bertini L (2007) *268*
Greco C, see Bertini L (2007) *268*
Grepioni F, see Braga D (2005) *254*: 71–94

Grimme S, see Daasbjerg K (2006) *263*: 39–70

Grozema FC, Siebbeles LDA, Gelinck GH, Warman JM (2005) The Opto-Electronic Properties of Isolated Phenylenevinylene Molecular Wires. *257*: 135–164

Guiseppi-Elie A, Lingerfelt L (2005) Impedimetric Detection of DNA Hybridization: Towards Near-Patient DNA Diagnostics. *260*: 161–186

Di Giusto DA, King GC (2005) Special-Purpose Modifications and Immobilized Functional Nucleic Acids for Biomolecular Interactions. *261*: 131–168

Haase C, Seitz O (2007) Chemical Synthesis of Glycopeptides. *267*

Hansen SG, Skrydstrup T (2006) Modification of Amino Acids, Peptides, and Carbohydrates through Radical Chemistry. *264*: 135–162

Harry SR, see Carney CK (2007) *270*: 155–185

Heise C, Bier FF (2005) Immobilization of DNA on Microarrays. *261*: 1–25

Heitz V, see Collin J-P (2005) *262*: 29–62

Herrmann C, Reiher M (2007) First-Principles Approach to Vibrational Spectroscopy of Biomolecules. *268*

Higson SPJ, see Davis F (2005) *255*: 97–124

Hirayama N, see Sakai K (2007) *269*

Hirst AR, Smith DK (2005) Dendritic Gelators. *256*: 237–273

Holzwarth AR, see Balaban TS (2005) *258*: 1–38

Houk RJT, Tobey SL, Anslyn EV (2005) Abiotic Guanidinium Receptors for Anion Molecular Recognition and Sensing. *255*: 199–229

Huc I, see Brizard A (2005) *256*: 167–218

Ihmels H, Otto D (2005) Intercalation of Organic Dye Molecules into Double-Stranded DNA – General Principles and Recent Developments. *258*: 161–204

Imai H (2007) Self-Organized Formation of Hierarchical Structures. *270*: 43–72

Indelli MT, see Chiorboli C (2005) *257*: 63–102

Inoue Y, see Borovkov VV (2006) *265*: 89–146

Ishii A, Nakayama J (2005) Carbodithioic Acid Esters. *251*: 181–225

Ishii A, Nakayama J (2005) Carboselenothioic and Carbodiselenoic Acid Derivatives and Related Compounds. *251*: 227–246

Ishi-i T, Shinkai S (2005) Dye-Based Organogels: Stimuli-Responsive Soft Materials Based on One-Dimensional Self-Assembling Aromatic Dyes. *258*: 119–160

James DK, Tour JM (2005) Molecular Wires. *257*: 33–62

Jones W, see Trask AV (2005) *254*: 41–70

Justicia J, see Cuerva JM (2006) *264*: 63–92

Kambe N, see Fujiwara S-i (2005) *251*: 87–140

Kano N, Kawashima T (2005) Dithiocarboxylic Acid Salts of Group 1–17 Elements (Except for Carbon). *251*: 141–180

Kappe CO, see Kremsner JM (2006) *266*: 233–278

Kaptein B, see Kellogg RM (2007) *269*

Karlen SD, Garcia-Garibay MA (2005) Amphidynamic Crystals: Structural Blueprints for Molecular Machines. *262*: 179–227

Kato S, Niyomura O (2005) Group 1–17 Element (Except Carbon) Derivatives of Thio-, Seleno- and Telluro-Carboxylic Acids. *251*: 19–85

Kato S, see Niyomura O (2005) *251*: 1–12

Kato T, Mizoshita N, Moriyama M, Kitamura T (2005) Gelation of Liquid Crystals with Self-Assembled Fibers. *256:* 219–236
Kaul M, see Pilch DS (2005) *253:* 179–204
Kaupp G (2005) Organic Solid-State Reactions with 100% Yield. *254:* 95–183
Kawasaki T, see Okahata Y (2005) *260:* 57–75
Kawashima T, see Kano N (2005) *251:* 141–180
Kay ER, Leigh DA (2005) Hydrogen Bond-Assembled Synthetic Molecular Motors and Machines. *262:* 133–177
Kellogg RM , Kaptein B, Vries TR (2007) Dutch Resolution of Racemates and the Roles of Solid Solution Formation and Nucleation Inhibition. *269*
King GC, see Di Giusto DA (2005) *261:* 131–168
Kirchner B, see Thar J (2007) *268*
Kitamura T, see Kato T (2005) *256:* 219–236
Kniep R, Simon P (2007) Fluorapatite-Gelatine-Nanocomposites: Self-Organized Morphogenesis, Real Structure and Relations to Natural Hard Materials. *270:* 73–125
Komatsu K (2005) The Mechanochemical Solid-State Reaction of Fullerenes. *254:* 185–206
Kremsner JM, Stadler A, Kappe CO (2006) The Scale-Up of Microwave-Assisted Organic Synthesis. *266:* 233–278
Kriegisch V, Lambert C (2005) Self-Assembled Monolayers of Chromophores on Gold Surfaces. *258:* 257–313

Lacôte E, see Albert M (2006) *264:* 1–62
Lahav M, see Weissbuch I (2005) *259:* 123–165
Lambert C, see Kriegisch V (2005) *258:* 257–313
Lansiaux A, see Dias N (2005) *253:* 89–108
Larhed M, see Nilsson P (2006) *266:* 103–144
Larhed M, see Wannberg J (2006) *266:* 167–197
Larsson O, see Du Q (2005) *261:* 45–61
Leigh DA, Pérez EM (2006) Dynamic Chirality: Molecular Shuttles and Motors. *265:* 185–208
Leigh DA, see Kay ER (2005) *262:* 133–177
Leiserowitz L, see Weissbuch I (2005) *259:* 123–165
Lhoták P (2005) Anion Receptors Based on Calixarenes. *255:* 65–95
Li WP, Meyer LA, Anderson CJ (2005) Radiopharmaceuticals for Positron Emission Tomography Imaging of Somatostatin Receptor Positive Tumors. *252:* 179–192
Liang Z, see Du Q (2005) *261:* 45–61
Lingerfelt L, see Guiseppi-Elie A (2005) *260:* 161–186
Liu S (2005) 6-Hydrazinonicotinamide Derivatives as Bifunctional Coupling Agents for ^{99m}Tc-Labeling of Small Biomolecules. *252:* 117–153
Liu S, Robinson SP, Edwards DS (2005) Radiolabeled Integrin $\alpha_v\beta_3$ Antagonists as Radiopharmaceuticals for Tumor Radiotherapy. *252:* 193–216
Liu XY (2005) Gelation with Small Molecules: from Formation Mechanism to Nanostructure Architecture. *256:* 1–37
Luderer F, Walschus U (2005) Immobilization of Oligonucleotides for Biochemical Sensing by Self-Assembled Monolayers: Thiol-Organic Bonding on Gold and Silanization on Silica Surfaces. *260:* 37–56

Maeda K, Yashima E (2006) Dynamic Helical Structures: Detection and Amplification of Chirality. *265:* 47–88
Magnera TF, Michl J (2005) Altitudinal Surface-Mounted Molecular Rotors. *262:* 63–97
Maini L, see Braga D (2005) *254:* 71–94

Malacria M, see Albert M (2006) *264*: 1–62
Marquette CA, Blum LJ (2005) Beads Arraying and Beads Used in DNA Chips. *261*: 113–129
Mascini M, see Palchetti I (2005) *261*: 27–43
Matsumoto A (2005) Reactions of 1,3-Diene Compounds in the Crystalline State. *254*: 263–305
McGhee AM, Procter DJ (2006) Radical Chemistry on Solid Support. *264*: 93–134
Meyer B, Möller H (2007) Conformation of Glycopeptides and Glycoproteins. *267*
Meyer LA, see Li WP (2005) *252*: 179–192
Michl J, see Magnera TF (2005) *262*: 63–97
Milea JS, see Smith CL (2005) *261*: 63–90
Mizoshita N, see Kato T (2005) *256*: 219–236
Möller H, see Meyer B (2007) *267*
Moonen NNP, Flood AH, Fernández JM, Stoddart JF (2005) Towards a Rational Design of Molecular Switches and Sensors from their Basic Building Blocks. *262*: 99–132
Moriyama M, see Kato T (2005) *256*: 219–236
Murai T (2005) Thio-, Seleno-, Telluro-Amides. *251*: 247–272
Murakami H (2007) From Racemates to Single Enantiomers – Chiral Synthetic Drugs over the Recent 20 Years. *269*
Mutule I, see Suna E (2006) *266*: 49–101

Naka K (2007) Delayed Action of Synthetic Polymers for Controlled Mineralization of Calcium Carbonate. *271*: 119–154
Nakayama J, see Ishii A (2005) *251*: 181–225
Nakayama J, see Ishii A (2005) *251*: 227–246
Neese F, see Sinnecker S (2007) *268*
Nguyen GH, see Smith CL (2005) *261*: 63–90
Nicolau DV, Sawant PD (2005) Scanning Probe Microscopy Studies of Surface-Immobilised DNA/Oligonucleotide Molecules. *260*: 113–160
Nilsson P, Olofsson K, Larhed M (2006) Microwave-Assisted and Metal-Catalyzed Coupling Reactions. *266*: 103–144
Niyomura O, Kato S (2005) Chalcogenocarboxylic Acids. *251*: 1–12
Niyomura O, see Kato S (2005) *251*: 19–85
Nohira H, see Sakai K (2007) *269*

Oda R, see Brizard A (2005) *256*: 167–218
Okahata Y, Kawasaki T (2005) Preparation and Electron Conductivity of DNA-Aligned Cast and LB Films from DNA-Lipid Complexes. *260*: 57–75
Okamura T, see Ueyama N (2007) *271*: 155–193
Oller-López JL, see Cuerva JM (2006) *264*: 63–92
Olofsson K, see Nilsson P (2006) *266*: 103–144
Oltra JE, see Cuerva JM (2006) *264*: 63–92
Onoda A, see Ueyama N (2007) *271*: 155–193
Otto D, see Ihmels H (2005) *258*: 161–204

Palchetti I, Mascini M (2005) Electrochemical Adsorption Technique for Immobilization of Single-Stranded Oligonucleotides onto Carbon Screen-Printed Electrodes. *261*: 27–43
Pascal R, Boiteau L, Commeyras A (2005) From the Prebiotic Synthesis of α-Amino Acids Towards a Primitive Translation Apparatus for the Synthesis of Peptides. *259*: 69–122
Paulo A, see Santos I (2005) *252*: 45–84
Pérez EM, see Leigh DA (2006) *265*: 185–208

Pilch DS, Kaul M, Barbieri CM (2005) Ribosomal RNA Recognition by Aminoglycoside Antibiotics. *253*: 179–204
Pividori MI, Alegret S (2005) DNA Adsorption on Carbonaceous Materials. *260*: 1–36
Piwnica-Worms D, see Sharma V (2005) *252*: 155–178
Polito M, see Braga D (2005) *254*: 71–94
Poulin-Kerstien AT, see Dervan PB (2005) *253*: 1–31
Procter DJ, see McGhee AM (2006) *264*: 93–134

Quiclet-Sire B, Zard SZ (2006) The Degenerative Radical Transfer of Xanthates and Related Derivatives: An Unusually Powerful Tool for the Creation of Carbon–Carbon Bonds. *264*: 201–236

Ratner MA, see Weiss EA (2005) *257*: 103–133
Raymond KN, see Seeber G (2006) *265*: 147–184
Rebek Jr J, see Scarso A (2006) *265*: 1–46
Reckien W, see Thar J (2007) *268*
Reiher M, see Herrmann C (2007) *268*
Renaud P, see Darmency V (2006) *263*: 71–106
Robinson SP, see Liu S (2005) *252*: 193–216

Saha-Möller CR, see You C-C (2005) *258*: 39–82
Sakai K, Sakurai R, Hirayama N (2007) Molecular Mechanisms of Dielectrically Controlled Resolution (DCR). *269*
Sakai K, Sakurai R, Nohira H (2007) New Resolution Technologies Controlled by Chiral Discrimination Mechanisms. *269*
Sakamoto M (2005) Photochemical Aspects of Thiocarbonyl Compounds in the Solid-State. *254*: 207–232
Sakurai R, see Sakai K (2007) *269*
Sakurai R, see Sakai K (2007) *269*
Saladino R, Crestini C, Costanzo G, DiMauro E (2005) On the Prebiotic Synthesis of Nucleobases, Nucleotides, Oligonucleotides, Pre-RNA and Pre-DNA Molecules. *259*: 29–68
Santos I, Paulo A, Correia JDG (2005) Rhenium and Technetium Complexes Anchored by Phosphines and Scorpionates for Radiopharmaceutical Applications. *252*: 45–84
Santos M, see Szathmáry E (2005) *259*: 167–211
Sato K (2007) Inorganic-Organic Interfacial Interactions in Hydroxyapatite Mineralization Processes. *270*: 127–153
Sauvage J-P, see Collin J-P (2005) *262*: 29–62
Sawant PD, see Nicolau DV (2005) *260*: 113–160
Scandola F, see Chiorboli C (2005) *257*: 63–102
Scarso A, Rebek Jr J (2006) Chiral Spaces in Supramolecular Assemblies. *265*: 1–46
Scheffer JR, Xia W (2005) Asymmetric Induction in Organic Photochemistry via the Solid-State Ionic Chiral Auxiliary Approach. *254*: 233–262
Schenning APHJ, see Ajayaghosh A (2005) *258*: 83–118
Schmidtchen FP (2005) Artificial Host Molecules for the Sensing of Anions. *255*: 1–29 Author Index Volumes 251–255
Schindler J, see (2007) *269*
Schoof S, see Wolter F (2007) *267*
De Schryver F, see De Feyter S (2005) *258*: 205–255
Schulten K, see Dittrich M (2007) *268*

Seeber G, Tiedemann BEF, Raymond KN (2006) Supramolecular Chirality in Coordination Chemistry. 265: 147–184
Seitz O, see Haase C (2007) 267
Senn HM, Thiel W (2007) QM/MM Methods for Biological Systems. 268
Sewell SL, see Carney CK (2007) 270: 155–185
Sharma V, Piwnica-Worms D (2005) Monitoring Multidrug Resistance P-Glycoprotein Drug Transport Activity with Single-Photon-Emission Computed Tomography and Positron Emission Tomography Radiopharmaceuticals. 252: 155–178
Shinkai S, see Ishi-i T (2005) 258: 119–160
Sibi MP, see Zimmerman J (2006) 263: 107–162
Siebbeles LDA, see Grozema FC (2005) 257: 135–164
Silvi S, see Balzani V (2005) 262: 1–27
Simon P, see Kniep R (2007) 270: 73–125
Sinnecker S, Neese F (2007) Theoretical Bioinorganic Spectroscopy. 268
Skrydstrup T, see Hansen SG (2006) 264: 135–162
Smith CL, Milea JS, Nguyen GH (2005) Immobilization of Nucleic Acids Using Biotin-Strept(avidin) Systems. 261: 63–90
Smith DK, see Hirst AR (2005) 256: 237–273
Specker D, Wittmann V (2007) Synthesis and Application of Glycopeptide and Glycoprotein Mimetics. 267
Stadler A, see Kremsner JM (2006) 266: 233–278
Stibor I, Zlatušková P (2005) Chiral Recognition of Anions. 255: 31–63
Stoddart JF, see Moonen NNP (2005) 262: 99–132
Strauss CR, Varma RS (2006) Microwaves in Green and Sustainable Chemistry. 266: 199–231
Suk D-H, see Crich D (2006) 263: 1–38
Suksai C, Tuntulani T (2005) Chromogenetic Anion Sensors. 255: 163–198
Sun J-S, see Escudé C (2005) 253: 109–148
Suna E, Mutule I (2006) Microwave-assisted Heterocyclic Chemistry. 266: 49–101
Süssmuth RD, see Wolter F (2007) 267
Svith H, see Daasbjerg K (2006) 263: 39–70
Swerdlow H, see Du Q (2005) 261: 45–61
Szathmáry E, Santos M, Fernando C (2005) Evolutionary Potential and Requirements for Minimal Protocells. 259: 167–211

Taira S, see Yokoyama K (2005) 261: 91–112
Takahashi H, see Tamura R (2007) 269
Takahashi K, see Ueyama N (2007) 271: 155–193
Tamiaki H, see Balaban TS (2005) 258: 1–38
Tamura R, Takahashi H, Fujimoto D, Ushio T (2007) Mechanism and Scope of Preferential Enrichment, a Symmetry Breaking Enantiomeric Resolution Phenomenon. 269
Thar J, Reckien W, Kirchner B (2007) Car-Parrinello Molecular Dynamics Simulations and Biological Systems. 268
Thayer DA, Wong C-H (2007) Enzymatic Synthesis of Glycopeptides and Glycoproteins. 267
Thiel W, see Senn HM (2007) 268
Tiedemann BEF, see Seeber G (2006) 265: 147–184
Tobey SL, see Houk RJT (2005) 255: 199–229
Toda F (2005) Thermal and Photochemical Reactions in the Solid-State. 254: 1–40
Tour JM, see James DK (2005) 257: 33–62
Trask AV, Jones W (2005) Crystal Engineering of Organic Cocrystals by the Solid-State Grinding Approach. 254: 41–70

Tuntulani T, see Suksai C (2005) *255*: 163–198

Uccelli L, see Boschi A (2005) *252*: 85–115
Ueyama N, Takahashi K, Onoda A, Okamura T, Yamamoto H (2007) Inorganic–Organic Calcium Carbonate Composite of Synthetic Polymer Ligands with an Intramolecular NH···O Hydrogen Bond. *271*: 155–193
Ushio T, see Tamura R (2007) *269*

Varma RS, see Strauss CR (2006) *266*: 199–231
Veciana J, see Amabilino DB (2006) *265*: 253–302
Venturi M, see Balzani V (2005) *262*: 1–27
Vezin H, see Dias N (2005) *253*: 89–108
Vögtle F, see Fages F (2005) *256*: 77–131
Vögtle M, see Žinić M (2005) *256*: 39–76
Volkmer D, see Fricke M (2007) *270*: 1–41
Vries TR, see Kellogg RM (2007) *269*

Walschus U, see Luderer F (2005) *260*: 37–56
Walton JC (2006) Unusual Radical Cyclisations. *264*: 163–200
Wannberg J, Ersmark K, Larhed M (2006) Microwave-Accelerated Synthesis of Protease Inhibitors. *266*: 167–197
Warman JM, see Grozema FC (2005) *257*: 135–164
Warren JD, Geng X, Danishefsky SJ (2007) Synthetic Glycopeptide-Based Vaccines. *267*
Wasielewski MR, see Weiss EA (2005) *257*: 103–133
Weiss EA, Wasielewski MR, Ratner MA (2005) Molecules as Wires: Molecule-Assisted Movement of Charge and Energy. *257*: 103–133
Weissbuch I, Leiserowitz L, Lahav M (2005) Stochastic "Mirror Symmetry Breaking" via Self-Assembly, Reactivity and Amplification of Chirality: Relevance to Abiotic Conditions. *259*: 123–165
Williams LD (2005) Between Objectivity and Whim: Nucleic Acid Structural Biology. *253*: 77–88
Wittmann V, see Specker D (2007) *267*
Wright DW, see Carney CK (2007) *270*: 155–185
Wolter F, Schoof S, Süssmuth RD (2007) Synopsis of Structural, Biosynthetic, and Chemical Aspects of Glycopeptide Antibiotics. *267*
Wong C-H, see Thayer DA (2007) *267*
Wong KM-C, see Yam VW-W (2005) *257*: 1–32
Würthner F, see You C-C (2005) *258*: 39–82

Xia W, see Scheffer JR (2005) *254*: 233–262

Yam VW-W, Wong KM-C (2005) Luminescent Molecular Rods – Transition-Metal Alkynyl Complexes. *257*: 1–32
Yamamoto H, see Ueyama N (2007) *271*: 155–193
Yashima E, see Maeda K (2006) *265*: 47–88
Yokoyama K, Taira S (2005) Self-Assembly DNA-Conjugated Polymer for DNA Immobilization on Chip. *261*: 91–112
Yoshikawa I, see Araki K (2005) *256*: 133–165
Yoshioka R (2007) Racemization, Optical Resolution, and Crystallization-induced Asymmetric Transformation of Amino Acids and Pharmaceutical Intermediates. *269*

You C-C, Dobrawa R, Saha-Möller CR, Würthner F (2005) Metallosupramolecular Dye Assemblies. *258*: 39–82
Yu J, see Dittrich M (2007) *268*
Yu S-H (2007) Bio-inspired Crystal Growth by Synthetic Templates. *271*: 79–118

Zampella G, see Bertini L (2007) *268*
Zard SZ, see Quiclet-Sire B (2006) *264*: 201–236
Zhang W (2006) Microwave-Enhanced High-Speed Fluorous Synthesis. *266*: 145–166
Zhang X-E, Deng J-Y (2005) Detection of Mutations in Rifampin-Resistant *Mycobacterium Tuberculosis* by Short Oligonucleotide Ligation Assay on DNA Chips (SOLAC). *261*: 169–190
Zimmerman J, Sibi MP (2006) Enantioselective Radical Reactions. *263*: 107–162
Žinić M, see Fages F (2005) *256*: 77–131
Žinić M, Vögtle F, Fages F (2005) Cholesterol-Based Gelators. *256*: 39–76
Zipse H (2006) Radical Stability—A Theoretical Perspective. *263*: 163–190
Zlatušková P, see Stibor I (2005) *255*: 31–63

Subject Index

Abalone shell, nacre 121
ACC 8, 38, 123, 125
Additive adsorption, face-selective 10
Amide NH, prelocated 155, 161
Amorphous calcium carbonate (ACC) 8, 38, 123, 125
Anatase 12
Antifreeze proteins 24
Aragonite 18, 26, 128, 174
Arylsulfatase 163
Asp-oligopeptide 185
Asprich proteins 23
Au nanoparticles, stabilization 45

BaCrO$_4$, DHBCs 50
BaSO$_4$, DHBCs 50
Bio-inspired mineralization 1
Biomineralization, hydrophilic polymers 23
Biopolymers 26, 85
–, matrix 103
Bis(2-ethylhexyl)sodium sulfate (AOT) 126
2,6-Bis(acetylamino)benzoate 163
Block copolymers 90

Ca 156
Ca cluster, synthetic chelating ligand 173
CaCO$_3$ composite, polymer ligand tacticity 180
–, crystal surface 176
–, helix 20
–, nanoparticles, amorphous 19
–, thin films, matrix-mediated formation 131
CaCO$_3$/poly(acrylate) 178
CaCO$_3$/poly[(Z)-3-allylcarbamoyl-2-propenate] 187

CaCO$_3$/polycarboxylate 174
CaCO$_3$/stearate monolayer 178
Calcite 128, 174
Calcium carbonate 123
–, crystallization, synthetic substrates 133
–, precipitation 123
–, surface, template mineralization 147
Calcium tartrate tetrahydrate (CaT) 48
Ca–O bond, carboxylate/phosphate Ca(II) complexes 164
Carbon nanotubes 3
Carboxylate Ca(II) complexes, NH-O hydrogen bond 167
CBP1 10
Cd–S bond 159
CdS, PEO-b-PEI 44
Celestine, peanut shaped 33
Cellulose 175
Chitin 121, 158
– fibers 174
Chondroin-4-sulfate 24
Chondroin-6-sulfate 24
Chondroitin-4-sulfate 26
Cobalt oxalate dihydrate 16
Collagen 26
Conformation switch, proton-driven 185
Copper oxalate 16
Crassostrea gigas 130
Crayfish, calcium ion storage 158
Crystal growth 79, 81
–, modifiers 96
–, synthetic template controlled 84
Crystal shape 81
Crystallization 79
–, foreign external templates 107
–, mixture of solvents 98
–, modes/modification 5
–, pathways 7

–, patterned surfaces 109
–, polymer-controlled 22

Delay addition 120
Dendrimers 38, 95, 126, 174
Dentin phosphoryns 23
Deoxyribonuclease I 161
Dermatan sulfate 24
Dextran 26, 27
DHBCs 1, 25, 39
DHGCs 25, 64
N,N-Dimethyl-S –carbamates 139
DNA 27
Double hydrophilic block copolymers (DHBCs) 1, 25, 39
Double hydrophilic graft copolymers (DHGCs) 25, 64

Ecdysis, crayfish 158

Fe 156
Fluoroapatite, crystallization 26
Foldamers 96

Gastroliths 158
Gastropod nacre 18
Glycoaminoglycans (GAGs) 24
Glycoproteins 24, 174
Gold colloids, mercaptophenol-protected 174
Gold nanoparticles, stabilization 45
Graft copolymers 90

Haliotis laevigata 18
Hashemite 50
Hematite, PMVE-*b*-PVOBA 43
Heparin 24
Heparin sulfate 24
Hexacyclen 45
5-Hexadecyloxyisophthalic acid 128
Homopolymers 29
Hydroxyapatite (HAP) 158
–, chondroitin-4-sulfate 26
–, fibers 13
Hydroxymethylpropylcellulose (HPMC) 16

Ice inhibition proteins 24
In situ polymerization 120
Inorganic–organic building blocks 80

Interface 79

K_2SO_4–PAA 36
Kemp's acid, amidated 186
Keratin sulfate, hyaluronic acid 24

Latent inductor 120
Lectin 24
Lingula anatina shells 26

Mammillan 24
Matrix 79
Mercenaria mercenaria 130
Mesocrystals 1, 14
Mesoscopic transformation 5
Metal oxo clusters 120
Metal–O bonding, carboxylate ligands 184
Mg 156
Mineralization, bio-inspired 25
Mn 156
Monolayers, interfaces 99
Mucoperlin 24

Nacre 18, 19
Nanocomposite, nacreous layer 183
Non-classical crystallization 1
Nucleation, $CaCO_3$, delayed action 132
–, multi-step, mineralization 129

Oligosilsesquioxanes 120
Oriented attachment 1, 12
Ovoglycan 24

P450 porphinate 160
PHEMA hydrogel replica, sea urchin spine 30
Phosphoinositide 161
Pinctada fucata, nacreous layer 156, 183
Pinna nobilis nacre 24
pKa shift 161
Poly(acrylate) 174, 175
Poly(acrylic acid), delay addition 133
Poly(allylamine hydrochloride) (PAH), $BaCO_3$ superstructures 33
Poly(Asp)
Poly(Glu) 174
Poly(isocyanide), chiral 174
Poly(propylene imine) dendrimers, octadecylamine 38

Subject Index

Poly(sodium 4-styrenesulfonate) (PSS)   33
Poly(styrenesulfonate)   126
Poly[3-(2,2,2-triphenylacetylamino)propionic acid]   180
Poly(vinylalcohol)   126
Polyamidoamine dendrimers, delay addition   139
Polycarboxylates   33, 176
Polyelectrolytes   86
Polymer induced liquid-precursor (PILP) process   20
Polymer ligands, biological relevance   182
–, tacticity   180
Polymers, responsive to stimulation   143
Porphyrin-template monolayer   174
PVA shells   38

Random copolymers   29

Sea urchins spines   18
Self-assembly   1, 79
Si   156
Sialoproteins   23

Silaffins   28
Solution–precursor–solid (SPS) mechanism   30
Sponge spikes   18
Statherin   184
Stearic acid monolayer   174
Sulfonate Ca(II) complexes, NH-O hydrogen bond   166
Supramolecular functional polymer   96
Surfactant   126
–, low mass   97

Tb(III) luminescence   163
Terbium   163
Transformation   120

Vaterite   126, 174
–, spherulitic   20

Witherite   58
Wulff rule   7

Zeolite   120
ZnO, stack of pancake   46

Printing: Krips bv, Meppel
Binding: Stürtz, Würzburg